OCEAN RESOURCES:
AN INTRODUCTION TO ECONOMIC OCEANOGRAPHY

Roger Henri Charlier
Bernard Ludwig Gordon

with contributions by Michel Vigneaux and Joseph Gordon

University Press of America™

GC
1015.2
.C3851

ENGINEERING
& PHYSICS
LIBRARY

Library of Congress Catalog Card Number: 78-61393

OTHER BOOKS BY R. H. CHARLIER

    INTRODUCTORY EARTH SCIENCE (1961)
    THE PHYSICAL ENVIRONMENT (1967)
    THE STUDY OF THE OCEANS (1969)
    THE STUDY OF ROCKS (1969)
    HARNESSING THE ENERGIES OF THE OCEAN (1970)
    ECONOMIC OCEANOGRAPHY (1976)
    TIDAL POWER (In Press)

OTHER BOOKS BY B. L. GORDON

    THE MARINE FISHES OF RHODE ISLAND (1960)
    MAN AND THE SEA (1970)
    MARINE CAREERS (1970)
    MARINE RESOURCE PERSPECTIVES (1974)
    HURRICANE IN SOUTHERN NEW ENGLAND (1976)
    ENERGY FROM THE SEA (1977)
    THE SECRET LIVES OF FISHES (1977)

For all our students

past, present, and future

Acknowledgements

Appreciation is expressed to Professor Michel Vigneaux, Director of the Institut de Géologie du Bassin d'Aquitaine of the Universite de Bordeaux, for graciously contributing to the content of the book, and to Joseph Gordon, former Navy officer, biologist and educator for his motivating influence, his many suggestions, and particularly for guiding this volume through publication. Gratitude is also duly expressed to the many government and private agencies and individuals who contributed the basic facts and figures without which the preparation of this text would have been impossible. And last, but not least, the authors also express their thanks to Joanne Kushner for her expertise and diligence in typing the manuscript and meeting all deadlines so that publication could proceed.

R.H.C.

B.L.G.

V

CONTENTS

CONTENTS

CONTENTS

CONTENTS

LIST OF FIGURES

LIST OF TABLES

PART ONE

THE SCOPE OF GEOGRAPHY

I

THE DICHOTOMOUS CHARACTER OF GEOGRAPHY

GEOGRAPHY concentrates on the study of regions or characteristics
of regions. Regional geography is the in-depth study of an area
or a region; it is basically a synthesizing discipline. The definition of a region, its nature, and the various types have been
the subject of much study and many monographs have been published
among which Robert Hartshorne's The Nature of Geography (Washington, Association of American Geographers) and Jan Broeck's The
Scope of Geography are milestones. How the concept of region can
perhaps be extended to the oceanic areas as well, has been the
subject of a recent article by Roger Charlier: "L'océanographic
régionale, discipline de synthèse" (V.U.B. Uitgaven, Brussel,
1975, Acta van het Interfacultair Colloquium over Multidiscipli-
nareit Interdisciplinareit en Trandisciplinariteit in Mens en
Natuurwetenschappen, pp. 207-218).

Most earth regions have characteristics belonging to two
groups of features: (a) the physical features, part of the original physical environment, and (b) the so-called cultural features, which have been superimposed on the physical ones by man
through his presence and his utilization of the available resources. Nature provides the geographical scene with organic and
inorganic elements: their study is the province of physical geography, and its corollary, biological geography. Human, cultural,
or perhaps better social geography, studies people and material
culture features. It is thus obvious that geography, the science
dealing with the features of the earth surface and of spacial distribution, is, by its very object, dichotomous, belonging simultaneously to the realms of both the "natural" and the "social"
sciences. Geography aims to systematically describe and interpret
the distribution of "features" or "things" on the surface of the
earth. The term "surface" must be taken sensu largo, because the
geographer includes in his sphere of concern the air above the
earth's surface, and the lithosphere underneath it. Though neglected to too large an extent in the past, the oceans, representing some 71% of the earth's surface are part of the geographer's
domain.

Summarizing, the geographer studies a thin zone that includes

3

the atmosphere-surface contact and a solid and liquid zone below
the surface wherein inorganic and organic forms intermingle and
interrelate.

1. PHYSICAL FEATURES

The physical features of a region include the very surface
on which man lives, thus the soil, arable or not, and the rocky
substratum. Rocks arrange themselves in a wide variety of con-
figurations and result from several possible origins, recently
formed in surface, formed underneath the surface, now soft, then
hard; the surface topography and the elevations vary widely in
form, color and range.

Other physical features include climates, surface, and subsurface waters, and living plants and animals. Let it be noted
that rocks, soils, native plants and animals constitute economic
resources susceptible to economic exploitation. The study of the
physical features is often science with a humanized perspective,
because they are frequently considered in the light of their resource potentialities for human use and development. Original,
natural, not human-modified, features of a region are called the
"fundament".

1.1 LANDFORMS

The geographer must grasp an understanding first of the nature and origin of the rocky substratum, and next, of the forms
of occurrence of these earth materials or relief. Relief forms
fall within four broad classes: plains, plateaus, hills and
mountains, though there are surface features of lesser magnitude.
A general classification of landforms might consider three or-
ders: continental platforms and ocean basins would fall in the
first; the second order would embrace constructional landforms,
viz. those formed by endogenic forces, or forces originating
within the earth such as folding, faulting, warping, sea floor
spreading and volcanism, resulting in plains, plateaus and various types of mountains; finally, the third order results from
exogenic forces, or the action of forces at the surface of the
earth, such as wind, water, ice, sun, gravity leading to either
erosion or deposition: mountains, plateaus, valleys, fans, del-
tas, beaches, moraines, dunes, to name a few, are thus created.

1.2 CLIMATE

The geographer is interested in both the characteristics

and the distribution of the world's climates. This requires
knowledge of maximal and minimal daily, monthly and yearly temperatures, as well as their ranges, and the extent of the frost-
free season. Precipitations are a major concern as well: their
amount, their frequency and their reliability, and their distribution throughout the year. Finally, atmospheric dynamics are
important, yielding information as to air masses, and types of
winds.

1.3 WATER

Surface and subsurface waters, and seas and oceans not only
play a considerable role in shaping the natural features of a
region, but also in determining human establishment. Drainage
patterns and river regimes, permanence of natural waterways and
frequency of floods, even water chemistry, are part of the physical make-up of a region. Waters often are the sculptor of surface and subsurface landforms: karst topography, caverns and
grottoes, for instance. Stream erosion was instrumental in low-
ering the Alps some 500 meters, and in transforming the Flanders
sand and line plateau into a peneplain.

1.4 SOILS

Soils are of crucial importance for economic geography, one
of the social geography disciplines. They must be known for their
physical and chemical properties and their profile. Distribution
of soil types plan an important role in human occupation. Soil
and water are the two most necessary earth resources. They play
a primary and continuing role in the distribution of plant and
animal life.

1.5 BIOTIC COVER

Biological geography is concerned with the distribution and
types of plants and animals. Forest, grass, shrubs, or the lack
thereof, determine the occurrence of animals. Plant and animal
life constitute in many cases important economic resources. Likewise, the ocean and its life forms are of great economic significance.

2. CULTURAL FEATURES

Besides the features readily provided a region by nature,
man has added numerous so-called "cultural" features by concen-
trating in some and not in others, by organizing settlements

according to given patterns, by adopting specific types of housing, by establishing farms, building factories, digging mines,
and constructing means of communication and transportation. To
these tangible modifications, one may add other cultural features:
type of life, political organization, moral and religious life,
education, racial and ethnic patterns. Nevertheless, the major
data of social geography concerns itself with forms of land utilization and ensuing trade patterns.

2.1   HUMAN OCCUPANCE

Excluding "life" type (French: "genre de vie") and social
organization, traditionally labeled anthropogeography and political geography, settlement geography is a weighty factor for
the economic development of a region which requires a labor pool.
Hence the geographer will study population density and distribution patterns. The arithmetic density of a population is the
ratio of the total population of a given area to its superficies
and constitutes an excellent indicator of concentration or dis-
persion patterns of distribution. Geographers occasionally also
call upon the physiological density in which they substitute
arable land for the total superficies, or the agricultural density which is the ratio of agricultural people to the arable land.
Pre-World War II Germany, for instance, had a total population of
69,486,000, and arithmetic, physiological and agricultural den-
sities of 371.9, 793 and 125. While the arithmetic density for
the post-war Federal Republic has increased, the agricultural
density has decreased.

Shelter study also includes house type, configuration, construction material, site and situation.

The social sciences aspect of geography has been labeled by
some geographers the "Geography of the Human Communities". They
see social anthropology and its connecting disciplines of economics, political science and sociology as the contributing sciences, providing contemporary as well as historical observations.
The definition has some drawbacks since man is not part of all
the regions he exploits. Economic geography embraces the geography of production, the geography of transport, the geography
of trade, the geography of communications; it being understood
that production includes all types of production--agriculture,
husbandry, mining, manfacturing--as well as services. Economic
geography is also a geography of distribution of resources, of
exploitation, and of commerce.

2.2   RESOURCES AND EXPLOITATION

According to Finch and Trewartha, "in no sense is the physical surface to be thought of as the sole cause of the character
and distribution of material culture. It is the human group,
with its particular heritage of racial [and ethnic] endowments,
customs, habits, and training, that creates the features of land
utilization, and it is the human element also that largely deter-
mines their character and distribution. Physical conditions set
up only certain very flexible limits to land use". The study of
resources, their distribution, exploitation and trade, has traditionally been designated economic geography.

One may start with a review of the natural resources of a
region. This would include the occurrence, accessibility and
economic value of minerals, native plant and animal life. A logical follow-up would cover agriculture and animal husbandry, type
of agriculture and distribution pattern of arable land, size and
layout of fields and farms, crops, domestic animals. Logging,
fishing, hunting, trapping are exploitation of the natural resources.

Industrial exploitation is extractive; mining, in fact, is
an exploitation of a natural resource, so is manufacturing. Its
study embraces raw products, finished products, power resources
and the characteristics of the plant itself.

Goods must be brought to the consumer, and thus economic
geography considers communications means and transportation, the
routes with their density and patterns, the carriers, and foreign
and domestic trade as well as transit.

Man's economic activities, his efforts to produce, to transport and to market fall thus within the scope of economic geography. A basic knowledge of the land's physical features is
necessary to assess the occurrence and exploitation possibility
of a region's resources. Production covers agriculture and the
animal and mineral kingdoms. The agricultural potential of a
region depends on the distribution of cultivated land and its
relative importance in the overall economic picture. Does the
area produce for its own consumption or export? Is import of
foodstuffs necessary? Is agriculture of the intensive or extensive type? What percentage of the population is engaged in agricultural work? Produce itself is divided into four categories:
(1) cereals and animal foods; (2) fruits; (3) industrial crops;
(4) forests. Animal products are either food, clothing or industrial. Mineral products are in the hundreds, but exploitation
is governed by demand, profitability and location.

II

THE GEOGRAPHICAL APPROACH

THE GEOGRAPHER first systematically observes and records on maps and graphs the data he has gathered. The features he studies are far-flung, often world-wide. Thereafter he must look for an explanation. No limit is placed upon the kind of explanation. And as for the geographical facts, explanations belong to both the domains of the physical and the social sciences.

III

THE GLOBAL OCEAN AS A REGION

GEOGRAPHERS and geologists alike have concentrated their efforts on the study of the land, even though it represents only one fourth of the earth's surface. Doubtlessly this is due to a matter of accessibility and instrumentation. Today, however, technology has conquered the ocean domain and exploitation of the ocean is more a matter of timing and international politics than of capability. The economic potential of the ocean has been described as a "trillion dollar opportunity", hence a study of the economic geography of the ocean must be timely. (Fig. 1) The ocean is therefore the area selected for this course. The physical features will be limited only to some basic ideas. (Fig. 2)

1. THE RELIEF ZONES OF THE OCEAN

The shoreline is not the end of the continental mass, it is merely a contact zone between water and land. The continent continues for a more or less long distance underneath the water cover. This gently sloping area is called the continental shelf and

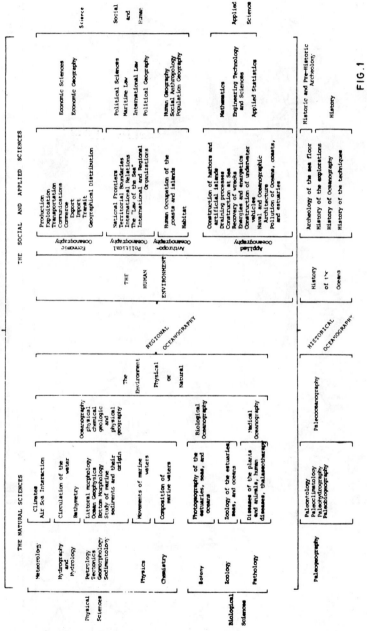

Oceanography: Study of the Marine Environment
SURVEYING COMPUTER SCIENCES CARTOGRAPHY

FIG. 1

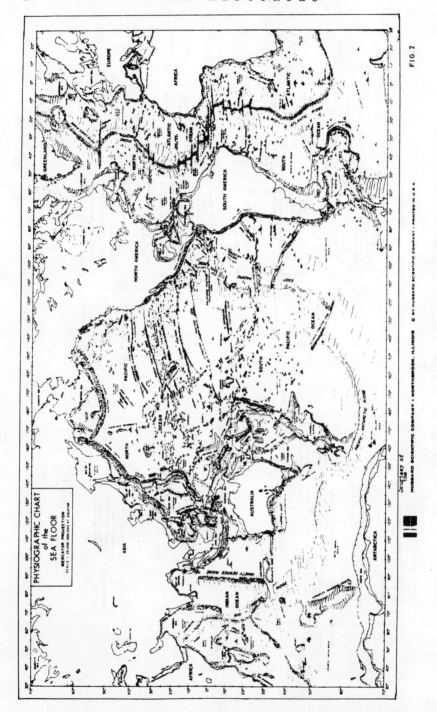

FIG. 2

Courtesy of

HUBBARD SCIENTIFIC COMPANY • NORTHBROOK, ILLINOIS   © BY HUBBARD SCIENTIFIC COMPANY • PRINTED IN U.S.A.

depths gradually increase to 200 meters. However, some shelfs
are very narrow and one reaches rapidly a break in the gradient,
that is where the continental slope starts, an area of lesser
biological activity, ending with a continental rise, itself followed by zones of great depths, viz. the abyssal and hadal zones.
Several topographic features are particular to the oceanic environment. At the water-land contact zone there are beaches or
cliffs, nearshore seacaves, stacks, spits, bars, reefs, lagoons,
and in the ocean itself sea-mounts, guyots and other forms.
(Fig. 3)

2. CLIMATES OF THE OCEAN

From a climatological point of view the ocean should be considered as made up of two major areas: the coastal ocean and the
open ocean. Oceanic and continental masses have reciprocal effects, but nowhere are they more pronounced than in coastal areas.
The climate zones of the ocean are based upon the similarity of
mean conditions at the water surface.

Four zones are recognized for the open ocean, though their
limits are somewhat arbitrary, generally convergence sites, where
the waters tend to accumulate and to sink and whose characteristics vary but little from one season to the next.

The coastal ocean waters are influenced by the conditions
prevailing upon the land. This, in turn, has an influence upon
the life in these areas. Often salinity is reduced near the
coast due to the debouching of rivers. As one gets further offshore, land influences decrease in impact, hence the near-shore
and off-shore coastal ocean climate zones.

3. THE NATURE OF OCEAN WATER

The composition of sea water is standard but additional materials may be found regionally and thus there is a geographical
variation in the relative concentration of the various chemicals
dissolved. Numerical results of water analysis are usually expressed as concentrations. Salinity is the amount of salts present in a kilogram of sea water. Previously this was determined
by chemical means. Today, oceanographers use either an electrical salinometer or an optical salinometer.

Marine currents affect sea water. They are caused by winds
and differences in density (e.g. salinity). They play an important role in life distribution, climate and navigation. Other
movements affect the ocean. Tides are primarily caused by lunar
and solar gravitational influences upon the earth. Tides and

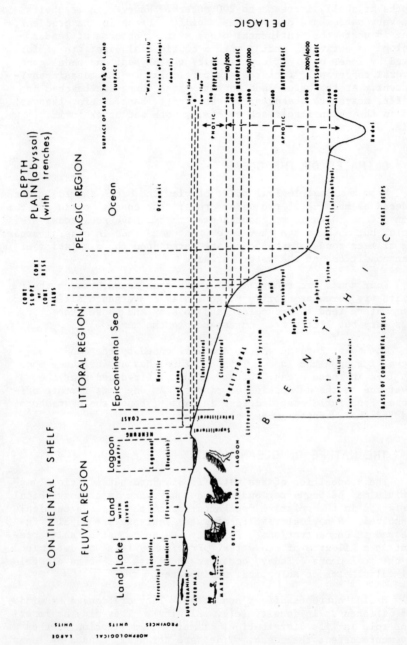

Fig. 3 ZONAL SUBDIVISIONS OF OCEANS

waves dissipate a considerable amount of energy. Insignificant
in some parts of the ocean, the amplitude of tides is considerable in some areas. Tidal currents carry alluvia, maintain channels and can be harnessed to generate electrical power.

Some water movements can be very destructive. The tsunami
produced by earthquakes and volcanic phenomena cause havoc on
coasts, while the relentless pounding of waves undermine cliffs
and often erode beaches, an action in which local currents often
play a role.

Ocean waters are also affected in some sites by vertical
movements. Surface water sinks (down welling) while cold deep
water loaded with nutrients rises to the surface (up welling).
The nutrients attract fish and the temperature differences of
upwelling and surface waters could be tapped as a source   of
energy.

The lithological nature of beaches varies from siliceous to
carbonaceous, from sands to gravel, and with the supply of construction materials dwindling on the land, offshore deposits become economically exploitable. Many rocks associated with ocean
floors are igneous but the abundance of sedimentary rocks of inorganic and organic origin in all parts of the ocean is the most
impressive. Ocean materials often have commercial value and mining many of them has started.

4. LIFE IN THE OCEAN

Plants and animals are found nearly everywhere in the ocean;
as on land, however, there are areas that may be considered biological deserts. The distribution of plant life is largely
governed by the possibility that photosynthesis can take place;
here depth and transparency of the water play a major role.

Solar radiation varies according to the ocean zones. It is
deeper in the neritic than in the oceanic zone. The neritic province corresponds roughly to the continental shelf relief zone
while the oceanic zone embraces slope, rise, abyssal and hadal
areas.

Oceanic life itself can be subdivided into two major types:
it is either pelagic or benthic; the former means it occupies
the waters, while the latter lives on the bottom or beneath it.
Both life zones and both life types are of economic importance.
Organic matter produced by marine plants, whether uni- or multi-
cellular, can be consumed by plant eating animals which in turn
are eaten by carnivorous animals, upon whom other carnivores

prey.   Man consumes plants and animals of the sea and through
aquaculture and husbandry attempts to increase his yield.   However, the low productivity of the ocean limits the role the seas
can play in human nutrition due in part to the conditions of
energy transfer from one echelon of the food or trophic pyramid
to the next, particularly to the terminal echelon of the benthic
pyramid.   The essential basis of the trophic pyramid is the primary production of photoautotrophic plants; however, some scientists believe in an important role of particulate aggregates
which associate dead organic matter and living matter represented
by the bacteria.   These aggregates represent a potential food
source for microphaga, especially filterfeeders, at the level of
the primary echelon unicellular organisms, and could enable secondary echelon animals to round off their plant food ration.
(Fig. 3B)

Pelagic life can be subdivided into two realms based upon
the ability of the organism to move in relationship to the water
masses. Plankton, which is made up of both plants and animals,
move up and down, and drift passively in the water mass transported with the currents. Plankton is the very basis of life in
the sea. (Figs. 4-7) All plankton are not protozoans and among
the larger forms are jellyfish and many mollusks. According to
size, plankton is subdivided into nannoplankton (5 to 60 microns),
microplankton (60 to 1000 microns) and macroplankton (over 1000
microns). Classified according to marine habitat, there is neritic or coastbound plankton and oceanic plankton. Plankton which
is tolerant of a wide range of ecological conditions and may be
found near shore and offshore is called panthalassic plankton.
Some animals live in a planktonic stage only during the early
part of their lives. They are designated as meroplankton. Zooplankton are animal plankton while phytoplankton are plant plankton. Nekton are swimmers. Most benthos or bottom dwellers possess locomotion power which permits bottom movement. This group
includes decopods and crustaceans, cephalopods, mollusks, and
worms.

Since all floating material is not alive, the term tripton
is used to designate such dead matter. Seston is the ensemble
of plankton and tripton.

Nekton occupies the upper echelons of the trophic pyramid
in the pelagic realm. It is, of course, inseparably linked to
plankton which occupies the lower echelons. Nektonic organisms
can travel from one water mass to another provided they can adapt
themselves physiologically to the different conditions of temperature and salinity or other factors. For some species these
migrations are governed by instinctive needs of the reproductive
processes. True nekton, like tuna, swims continuously in open

Fig. 3 B. Man's "harvest" from the ocean

Figure 4

NANNOPLANKTON

Nominal Size Range (Max. Dia.): 0.005–0.060 mm

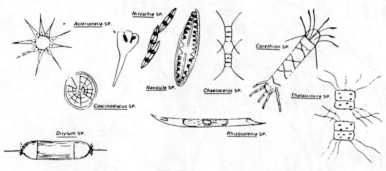

DIATOMS

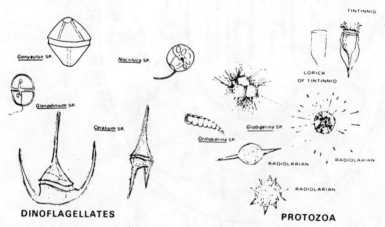

DINOFLAGELLATES   **PROTOZOA**

(Courtesy of MARTEK Instruments)

Figure 5
MICROPLANKTON
Nominal Size Range (Max. Dia.): 0.060–1.000 mm

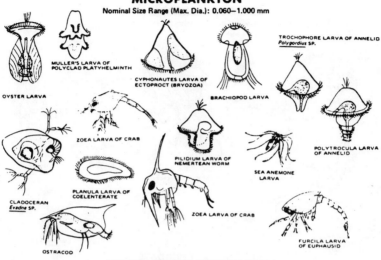

MULLER'S LARVA OF POLYCLAD PLATYHELMINTH

CYPHONAUTES LARVA OF ECTOPROCT (BRYOZOA)

TROCHOPHORE LARVA OF ANNELID *Polygordius* SP.

OYSTER LARVA

BRACHIOPOD LARVA

ZOEA LARVA OF CRAB

PILIDIUM LARVA OF NEMERTEAN WORM

POLYTROCULA LARVA OF ANNELID

SEA ANEMONE LARVA

CLADOCERAN *Evadne* SP.

PLANULA LARVA OF COELENTERATE

ZOEA LARVA OF CRAB

OSTRACOD

FURCILA LARVA OF EUPHAUSID

MISCELLANEOUS MICROPLANKTON

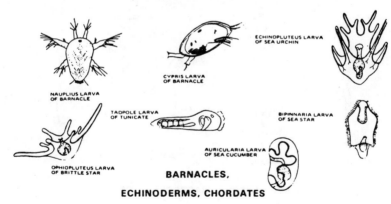

ECHINOPLUTEUS LARVA OF SEA URCHIN

CYPRIS LARVA OF BARNACLE

NAUPLIUS LARVA OF BARNACLE

TADPOLE LARVA OF TUNICATE

BIPINNARIA LARVA OF SEA STAR

AURICULARIA LARVA OF SEA CUCUMBER

OPHIOPLUTEUS LARVA OF BRITTLE STAR

BARNACLES, ECHINODERMS, CHORDATES

Figure 6

MACROPLANKTON
Nominal Size Range (Max. Dia.):
Larger than 1.000 mm

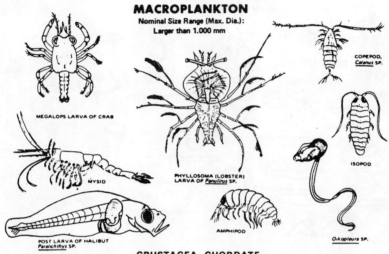

MEGALOPS LARVA OF CRAB

MYSID

PHYLLOSOMA (LOBSTER) LARVA OF *Panulirus* SP.

AMPHIPOD

POST LARVA OF HALIBUT *Paralichthys* SP.

COPEPOD, *Calanus* SP.

ISOPOD

Oikopleura SP.

CRUSTACEA, CHORDATE

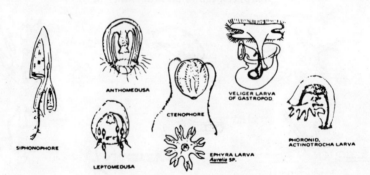

SIPHONOPHORE

ANTHOMEDUSA

LEPTOMEDUSA

CTENOPHORE

EPHYRA LARVA *Aurelia* SP.

VELIGER LARVA OF GASTROPOD

PHORONID, ACTINOTROCHA LARVA

MISCELLANEOUS MACROPLANKTON

Figure 7
MEGAPLANKTON

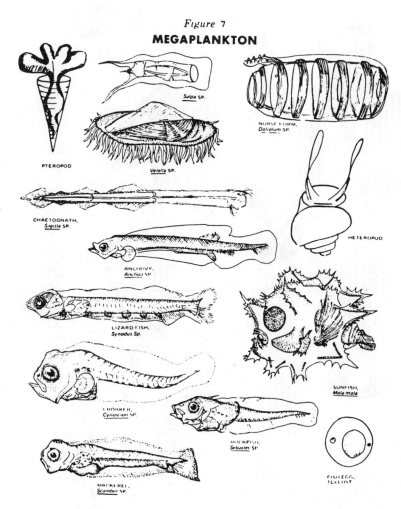

PTEROPOD

Salpa SP.

Velella SP.

NURSE FORM,
Doliolum SP.

CHAETOGNATH,
Sagitta SP.

HETEROPOD

ANCHOVY,
Anchoa SP.

LIZARD FISH,
Synodus Sp.

SUNFISH,
Mola mola

CROAKER,
Cynoscion SP.

ROCKFISH,
Sebastes SP.

MACKEREL,
Scomber SP.

FISH EGG,
TELEOST

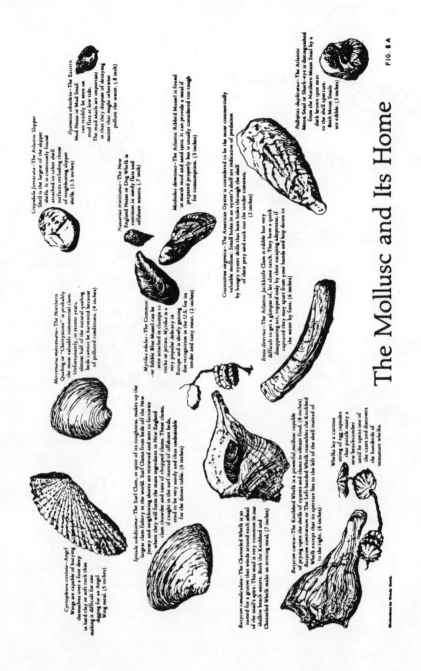

FIG. 8 A

The Mollusc and Its Home

Ilyanassa obsoleta—The Eastern Mud Nassa or Mud Snail can readily be seen on mud flats at low tide. The mud snails are important in that they dispose of decaying matter that might otherwise pollute the water. (.8 inch)

Crepidula fornicata—The Atlantic Slipper Shell is the largest of the slipper shells. It is commonly found attached to other shell surfaces including those of neighboring slipper shells. (1.5 inches)

Nassarius trivittatus—The New England Nassa or Dog Whelk is common in sandy flats and offshore waters. (.7 inch)

Modiolus demissus—The Atlantic Ribbed Mussel is found in marsh mud and sand spits. It can provide a meal if prepared properly but is usually considered too tough for consumption. (3 inches)

Polinices duplicatus—The Atlantic Moon Snail or Shark-eye is distinguished from the Northern Moon Snail by a dark brown spot next to the shell aperture. Both Moon Snails are edible. (3 inches)

Mercenaria mercenaria—The Northern Quahog or "Cherrystone" is probably the most valuable commercial clam. Unfortunately, in recent years, almost half of the natural quahog beds cannot be harvested because of polluted conditions. (4 inches)

Mytilus edulis—The Common or Edible Blue Mussel can be seen attached in clumps to rocks or jetties. Mytilus is a very popular delicacy in Europe and is slowly gaining due recognition in the U.S. for its tender and tasty meat. (2 inches)

Crassostrea virginica—The American Oyster is considered to be the most commercially valuable mollusc. Small holes in an oyster's shell are indicative of predation by hungry oyster drills that bore holes through the shell of their prey and suck out the tender contents. (3 inches)

Ensis directus—The Atlantic Jackknife Clam is edible but very difficult to get a glimpse of, let alone catch. They have a quick disappearing act, topped only by their escaping adeptness; if captured they may spurt from your hands and hop down to the water by foot. (6 inches)

Spisula solidissima—The Surf Clam, in spite of its toughness, makes up the larger clam fishery in the world. Surf Clams from beds off the New Jersey and neighboring shores are retrieved and sent to factories where they will form the main ingredient in New England clam chowder and cans of chopped clams. These clams, if caught in the surf instead of offshore beds, tend to be very sandy and thus undesirable for the dinner table. (6 inches)

Cyrtopleura costata—Angel Wings are capable of burying themselves over a foot deep in hard clay or soft rock thus making it difficult for one digging for an Angel Wing meal. (5 inches)

Busycon canaliculatum—The Channeled Whelk is so named for a groove that winds around each whorl of the snail's spire. This snail is very common on our shallow beach waters. Both the Knobbed and Channeled Whelk make an inviting meal. (7 inches)

Busycon carica—The Knobbed Whelk is a powerful mollusc capable of prying open the shells of oysters and clams to obtain food. (8 inches) *Busycon contrarium* or The Left-handed Whelk resembles the Knobbed Whelk except that its aperture lies to the left of the shell instead of to the right. (6 inches)

Whelks lay a curious string of egg capsules that contain many a new little creature until he opens one of the cases and discovers the hundreds of miniature whelks.

Illustration by Wendy Bink.

Littorina obtusata—The Northern Yellow Periwinkle's color often has a contrasting spiral stripe, perhaps yellow, against a reddish-brown background. (.3 inch)

Urosalpinx cinerea—The notorious Oyster Drill is an enemy to the oyster beds. As previously mentioned, the snail drills through the shell and sucks out his oyster meal. (.7 inch)

Littorina irrorata—The marsh is the home of the beautifully white and brown dotted Marsh Periwinkle, also an edible mollusc. (1 inch)

Littorina littorea—The Common European Periwinkle was originally a European species but has managed to spread to our North American coast. This periwinkle is a very popular food in European markets but once again Americans are slow to recognize a delicious and plentiful food source. (1 inch)

Donax fossor—Power Donax are very small clam shells that may be the most plentiful pelecypoda on the New Jersey coast. (.3 inch)

Astarte subaequilatera—Lentil Astarte is a bright red mollusc enclosed in a reddish-chestnut shell. (1 inch)

Divaricella quadrisulcata—The gray to whitish Cross-hatched Lucine is found in sandy areas. (.7 inch)

Crepidula convexa—The Convex Slipper Shell, like its relatives, is often attached to other shell surfaces. (.5 inch)

FIG. 8 B

Anadara transversa—The Transverse Ark is one of the Arks. These molluscs may be eaten but are not very popular because of their bitter tasting flesh. (1 inch)

Lunatia heros—The Northern Moon Snail is found in sand flats just below the low-tide mark or deeper. The Moon Snails lay their eggs on a distinctive sand collar which they have formed by cementing sand with mucous secreted from their mantle cavity. (4 inches)

Tagelus plebeius—or the Stout Tagelus is found in muddy, shallow areas. (2.5 inches)

Haminoea solitaria—The Eastern Paper Bubble carries a delicate white shell and lives in the intertidal zones. (.3 inch)

Anomia simplex—The Sawtooth or Plain Jingle Shell may be golden or silvery. (1 inch)

Eupleura caudata—The Thick-lipped Drill is also an abundant pest of oyster beds. (.7 inch)

Argopecten irradians—The Atlantic Bay Scallop is the most common edible scallop. Only the tender adductor muscle which holds the scallop's shell valves together is eaten. (3 inches)

Barnea truncata—The Fallen Angel Wing or Truncated Borer has a delicate white shell and like its relative, The Angel Wing, can bore deep into clay or mud. (2 inches)

Pandora gouldiana—Gould's Pandora is common in sandy and rocky areas, and in oyster beds. (1 inch)

Crepidula plana—The Eastern White Slipper Shell or The Flat Slipper Shell is often attached to the shells of whelks, snails and horseshoe crabs. Males are found attached to the shells of the much larger females. (1 inch)

Anadara ovalis—The Blood Ark has red blood, an unusual color for molluscs which tends to discourage prospective human consumers. (2 inches)

Petricola pholadiformis—The False Angel Wing or Rock Borer "bores" itself into rocks and clay. (2 inches)

Mya arenaria—Soft-shelled clam or "steamer" is one of the most important food molluscs. Several million dollars worth of these clams are harvested each year for steaming and chowder. (3 inches)

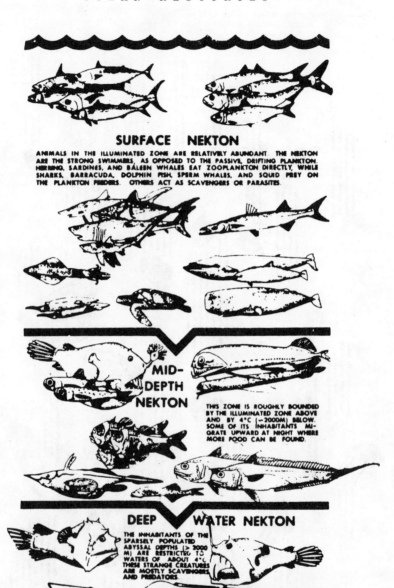

SURFACE NEKTON

ANIMALS IN THE ILLUMINATED ZONE ARE RELATIVELY ABUNDANT. THE NEKTON ARE THE STRONG SWIMMERS, AS OPPOSED TO THE PASSIVE, DRIFTING PLANKTON. HERRING, SARDINES, AND BALEEN WHALES EAT ZOOPLANKTON DIRECTLY, WHILE SHARKS, BARRACUDA, DOLPHIN FISH, SPERM WHALES, AND SQUID PREY ON THE PLANKTON FEEDERS. OTHERS ACT AS SCAVENGERS OR PARASITES.

MID-DEPTH NEKTON

THIS ZONE IS ROUGHLY BOUNDED BY THE ILLUMINATED ZONE ABOVE AND BY 4°C (~2000M) BELOW. SOME OF ITS INHABITANTS MIGRATE UPWARD AT NIGHT WHERE MORE FOOD CAN BE FOUND.

DEEP WATER NEKTON

THE INHABITANTS OF THE SPARSELY POPULATED ABYSSAL DEPTHS (> 2000 M) ARE RESTRICTED TO WATERS OF ABOUT 4°C. THESE STRANGE CREATURES ARE MOSTLY SCAVENGERS AND PREDATORS.

FIG 9

waters, but nektobenthos circulates near the bottom although not
bound to it.   (Fig. 9)

Benthos includes phyto and zoobenthos.   These organisms live
either above the bottom -- attached, walking, creeping -- swim
above it (nektobenthos) or exist underneath it.

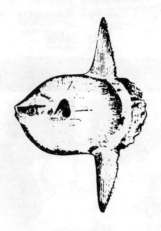

Mineral Harvesting

PART TWO

A STUDY OF OCEAN RESOURCES

"Knowledge of the ocean is more
than a matter of curiosity.
Our very survival may hinge on it."

JOHN F. KENNEDY

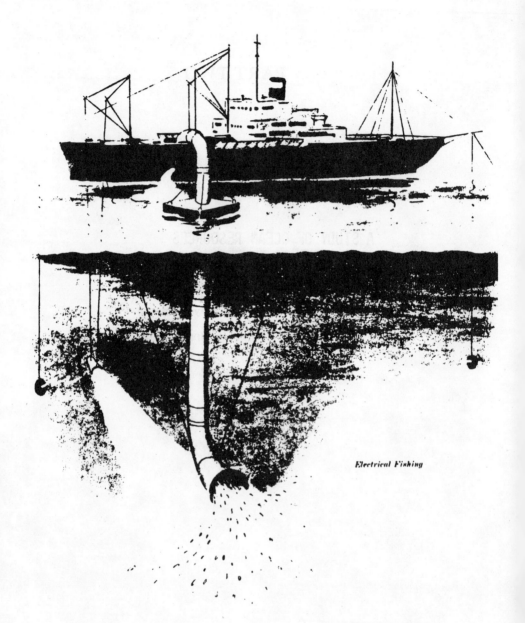

Electrical Fishing

TURN TO THE SEA

1. THE SEARCH FOR NEW RESOURCES

The needs of man are growing incessantly. This results from demographic growth, roughly 2% yearly or 10,000 additional humans per hour. Within 30 years, at this rate, there will be 7 1/2 billion humans, at least twice as many people as are on earth now. Another cause lies in the improvement of living standards: better transportation, more sophisticated foods, more comfort, more use of energy, to name a few. (Table 1)

Table 1. World Population (U.N.O. Data)

Area	Millions	% of World Total
Asia	2,200	57.0
Europe	472	12.2
U.S.S.R.	250	6.5
North America	236	6.1
South America	309	8.0
Africa	374	9.7
Oceania	20.6	0.5

Man will be forced to turn to the sea perhaps because there is no other more practical solution at this point in time. The ocean is a repository of most of the raw materials man needs, and perhaps technologists will find in the water domain nuclear energy, living space, and food. Yet, although exploitation of the ocean has barely begun, the spectre of marine pollution is becoming steadily more real. We must aim to use marine resources with care and caution with an eye toward the future.

Human and animal needs for fresh water have already become a major concern. Further demands are made by agriculture,

27

industry, tourism and recreation. St. Malo, where France's tidal
power station is located was threatened with water rationing in
1974, when water consumption rose, during the winter, from 5,900
m3 per day to 7,250 m3 per day. What would it become during the
summer when the population swells through a huge influx of tourists? Within 30 years, the need for fresh water and food will
have been doubled. Drought will increase mankind's use of water.
Man cannot afford to ravage the ocean as he did the earth. He
must adapt his desalinization methods and fishing techniques -
his harvesting methods - so that the ecosystem is not endangered.
Similarly, another challenge of the present decade is for the
marine "mining" engineer to develop exploitation methods and
techniques which will not constitute a danger to the marine environment.

An important question raised in official and private fields
alike is "Is there a catalyst that can bring together the maritime powers so that there is a rational management and an international management of marine technology in order to extract the
hoped for benefits of the oceans?". A first step in the right
direction might well be the preparation of an atlas of economic
oceanography, illustrating the resources of the ocean, showing
pollution sources, and picturing the circulation models of ocean
waters. Such an atlas would contribute to the protection of the
oceanic environment from man's continuous depredations. The
Soviets, for example, are ready to start a 200 vessel campaign
to draw the map of the economic wealth available in the 7 billion
square kilometers which constitute their national territorial
waters.

Edward Wenck wrote some time ago that marine technology
should be considered on a worldwide basis and not be limited to
the development of exploitation means of ocean organic and mineral products. This global image encompasses wastes disposal,
maritime trade, maintenance of peace, scientific research and
natural conservation. To this already lengthy list one should
add tourism, recreation, climate control , and the conservation
of energy to usable forms. Such an economic atlas is by no means
an idealist's dream. Today technologists can determine the geographic distribution of minerals located at depths exceeding 300
meters by using a nuclear probe containing man-made Californium-
252 followed by area scanning with a detector. Minerals, indeed,
absorb Californium-252, then emit gamma rays. Gold, silver,
copper, and manganese were thus detected, and the proponents of
this method claim that quantities as minute as 20 grams per ton
could thus be located.

Nuclear probe detection of minerals has been the object of
a research grant from the U. S. Atomic Energy Commission for

tests in Sequin Bay, Washington. This nuclear probe could be instrumental in drawing a map of the ocean bottom using geophysical
methods and when available will facilitate the evaluation of mineral deposits in various locations in a matter of minutes.

At times it appears that the viewpoints of the technologist
and of the champion of conservation are at odds. When this occurs should we not ask: "Is such opposition not reconcilable?"
Isn't there a path that can lead to a rational exploitation of
the Earth without running the risk of irremediable damage to the
environment and humanity? The promise that the oceans hold is
considerable and its resources may well be inexhaustible providing we use the sea carefully and wisely.

Often compared to a hidden treasure trove, the ocean keeps
in store many fields still veiled in mystery. How to ensure a
harmonious development of ocean utilization for peaceful purposes
is a goal present mankind must initiate and pass on to the future
generations. While there is a growing worldwide recognition that
resources deposited on and in the deep seabed are of vital importance to all humanity, the need to bring about a solution to
ocean pollution is becoming steadily more evident if biological
resources are to be safeguarded.

Looking at the rapidly dwindling stockpile of mineral resources on land, Boulding's suggestion some few years ago takes
on an urgent reality: "We can therefore regard the present period as a unique opportunity in the history of this particular
planet whereby the geological capital which has been accumulated
over hundreds of millions of years in the form of ores and fuels
can be spent to produce enough knowledge to enable man to do
without the geological capital which he exhausts."

It is true, however, that while offshore fuel reserves
already have proved themselves to be economic capital and their
extraction merely required extension of land   recovery   technology to   an   ever   deeper water milieu, hard mineral exploration and mining from the sea environment has, but in a few instances not yet proven to be economically enticing and does
require the development of a substantially new technology. Such
technology, however, according to all current indications presents no major difficulties.

2. EXPLOITATION OF THE OCEANS

Attracted by the probable wealth of the oceans, several
countries have set up scientific, technical and industrial
working organizations to first explore, then exploit the marine
resources. Nineteen nations have put more than 500 scientific

vessels into service, each searching for a share of the wealth.

Considerable progress has been made since the Challenger's 1872 voyage. The yearly spending of 600 million dollars for ocean research increases faster than inflation! Submersibles, once an exception, are now in common use. Since the start of the International Oceanographic Decade in 1970 sponsored by UNESCO to develop and coordinate marine research, Japan increased its budget fourfold, the German Federal Republic doubled its own, and the United States alone, spent yearly, over the first three years, $450 million dollars. The scramble for the ocean's riches has started. The budget of France's Centre National pour l'Exploitation des Océans (CNEXO) grew from $6.4 million in 1968 to $12 million in 1969 and $15 million for 1976.

2.1 VALUE OF OCEAN RESOURCES

The value of the fish catch, by 1980, is estimated at $15 billion a year, with 90% of all commercial fishing done within 200 miles from the shore. Offshore oil production will account for 33% of world oil production by 1980 and be worth at least $83 billion a year, with between 80% and 95% of such oil within 200 miles from the coast. By 1980 ocean shipping will be worth $51 billion per year, with most such commercial shipping within 200 miles from the coast. This totals $149 billion. Another $200 million dollars worth of manganese, copper, nickel and cobalt could be extracted yearly from ocean nodules and perhaps reach $1 billion by 1985. There are 22 billion tons of estimated nodules.

The contribution of the primary value of the fisheries to the gross national product, however, may in the future be of limited value since the increasing cost of oil may have consequences upon fish prices. Indeed modern long distance mechanized fishing is a top energy intensive food production industry, and ever more expensive synthetic materials for fishing gear are widely used. Furthermore, science conservation policy inadequacies in some cases have cost the world community hundreds of millions of dollars each year as seen by the decline of whaling and the Northwest Atlantic fisheries industry.

Recent computer studies indicate that an investment in exploitation of ocean resources will return in 20 years more than three times the same sum placed at 10% yearly interest. These studies considered the ocean as a source of food, an area of fishing, of aquaculture, mariculture, ostreiculture and conchyculture. Yet, the success of such exploitation is challenged by looming pollution. The studies also considered the ocean as a source of raw materials, and foresaw exploration, reconnaissance,

pre-exploitation planning, and extraction of petroleum, gas, and
ores.

The gigantic thermodynamic engine constituted by the ocean-
atmosphere complex is a primordial natural phenomenon and study
theme: ocean action upon meteorological conditions. The theme
"ocean, source of recreation and health" has gained considerable
momentum as humanity becomes steadily more concentrated in as-
phalted urban zones where green spots and sky glimpses are increasingly rare. Still another theme of current interest is to
find ways of transforming products extracted from the ocean into
consumer goods, and to ensure their marketing.

Oceanography is important to the consumer because of its
"products", and to the producers because of its technical needs.
"Oceanography is one of the most promising markets of the current
decade," stated J. Jamison Moore. "It offers a potential that
has unleashed the imagination and the enthusiasm of industrial
planners".

If, on the one hand, opportunities have been pictured as an
exceptional cornucopia, we cannot afford to lose sight of the
need to acquire first a thorough understanding of the environment. Today, a sluggish market supports only a limited effort,
except in petroleum and gas exploitation. A sober assessment of
priorities is imperative and expansion of existing markets must
be kept apart from the creation of new industries. The ocean
must be considered as a milieu subject to technological and ecological limitations. Reaping the ocean's resources will produce
a profit for the investor and improve the trends in national and
international economies.

2.2   TYPES AND AREAS OF EXPLOITATION

The California-based research institute Modern Management
recently undertook a study dealing with the various ways the
United States could make use of the ocean and reached the following conclusions:

(1) commercial activities, e.g. transportation, shipbuild-
ing, harbor exploitation, tourism and recreation: 81%;

(2) military activities, including submarine building and
the "Man-in-the-Sea" project: 8%;

(3) environmental problems, including education, research,
development, weather study and forecasting, draining, coastal
erosion, estuary management: 2%;

(4) chemical activities, e.g. desalination, pollution abatement, boring: 0.5%;

(5) biological activities, e.g., aquaculture, control of nocive or dangerous animals, fisheries, pharmacology: 4%;

(6) geological projects, inclusive of exploration for and exploitation of minerals, among which petroleum: 4 1/2%.

The study does not mention other activities as the physical exploitation of the ocean, e.g. tidal power plants, nor of the ocean's therapeutic use known as thalassotherapy.

Although geological activities appear to be rather modest, representing barely 4 1/2%, it should nevertheless be emphasized that outside of the communist countries, 80% of all the mineral reserves on land, are controlled by American capital, which explains the anxious interest on the part of American companies to extract the mineral resources from the oceans. According to a 1971 UNESCO report, already in 1969, mineral products valued globally at $7 billion were retrieved from the oceans.

In 1968, coastal locations accounted for 84% of all marine exploitation activities, 14% were offshore endeavors, and only 2% of the operations were deep water projects. The trend is rapidly changing and agreements signed in Geneva in 1958 which granted sovereign rights to riparian nations up to a depth of 200 meters, are now outdated since the ocean is readily exploitable beyond the 200 meter depth limit. In 1948 the first offshore well in Louisiana in the Gulf of Mexico reached a depth of 15 meters. By 1967, wells were being drilled in the ocean at nearly 400 meters. And it is unhesitantly being predicted that oil will be tapped, come 1980, at depths of some 2000 meters. With the new and increased demands of an expanding population, the economy of the ocean is a matter for immediate attention.

II

OCEAN RESOURCES

1. WATER

One of the resources of the ocean is water itself, whether it is used for the extraction of salts, or in hydroelectric plants, or for quenching the thirst of man and of parched lands.

1.1 LAVA COOLING

The 1973 eruption on the Icelandic island of Heimaey was the occasion for a major attempt to control a lava flow by the use of sea water as a cooling agent. Icelanders have coped with volcanic phenomena over many years. In the year 570 a previous eruption on this land was described by Brendan MacFinnloga, the Irish Abbot of Clonfert with particular awe.

Some 50 to 200 liters per second of sea water was poured onto the lava flow in the recent eruption. A wall of cool rubbly lava was thus created at the margin of the flow and the flow thickened against it. In places fractures which were originally deep in the flows were coated with salt left by the sea water that evaporated. While the island's fishing harbor and town of Vestmannaeyjar were heavily damaged by the eruption and its lava flows, the use of the cooling sea water limited further damage and was a significant development in using ocean water to modify or control the movement of lava.

1.2 A SOURCE OF FRESH WATER

The shortage of fresh water poses a pressing problem. Water levels are dropping on the continent. Occasionally fresh water wells are invaded by brackish waters. Artesian wells run dry. Deeper and deeper drilled wells are needed. Kings County, outside New York City, woke up one morning to find its wells put out of use by seepage of the Long Island Sound waters - this was a half a century ago. Today, New York City and the surrounding

communities get most of their drinking water from the Catskill
Mountains some 200 kilometers away (119 miles). Tapping the ocean
for fresh water, an unthinkable luxury not so long ago, is now a
necessity. Engineers have examined the possibility of towing
icebergs from Antarctica to Australia and to the west coast of
the United States. An iceberg with dimensions of 1,000 by 1,000
by 250 meters represents 250 million cubic meters and could be
towed in 300 days near the Atacama desert in Chile. It would lose
86% of its water mass but would still represent 35 million cubic
meters of fresh water worth 2.7 million dollars. The trip would
have cost 1.3 million dollars and thus, theoretically, the operation would be profitable. The fresh water available from Antarctic ice alone is staggering.

With 90% of the world's ice locked up in its cold storage,
Antarctica is easily considered the world's largest ice mass.
Snow and ice have been measured 15,000 meters thick at the South
Pole, with the average height of the ice sheet above sea level
estimated at 2,500 meters. In calculating the amount of water
locked in ice there, scientists have figured that if the cap
should melt, the oceans of the world would rise 260 meters, drown-
ing every coastline and wiping out every port, harbor and coastal
city.

Today, there are numerous desalting plants throughout the
world providing fresh water for man, cattle, and irrigation purposes. Salt removal is done through a variety of processes such
as distillation, the membrane process, freezing, solar distillation, chemical processes, physical and electrical methods. The
San Diego, California flash distillation unit was the first plant
in the United States to use the multiple flash process and is one
of the world's largest. The U. S. Office of Saline Water dis-
closed in 1968 that no less than 627 plants were then in use, being built, or on order, with a production capacity of 800,000
cubic meters per day. In many cases the energy needed to run the
plants is provided by petroleum or gas, ideal energy sources in
the Middle East. Kuwait is one of the leaders in fresh water
production producing 100,000 cubic meters per day and is building
two more plants which together will provide an additional 130,500
cubic meters of fresh water daily. Saudi Arabia is seriously at
work planning to utilize floating an iceberg from Antarctica to
help solve its fresh water problems.

The major problem, of course, is the cost of the produced
fresh water. Since fuel is no problem in Kuwait, water can be
produced at a cost of 12 cents per cubic meter; otherwise, it
would cost about 30 cents.

The price can probably be cut by planning larger plants,

treating at least 100,000 cubic meters per day, and establishing
agricultural-industrial complexes which will recuperate the waste
products. If plans to build nuclear powered plants go through,
then giant desalting factories will further reduce production
costs by furnishing between 300,000 and 500,000 cubic meters per
day.

1.3 THERAPEUTIC USES

Is salt water beneficial to health? Agreement was reached
long ago on the favorable effects of marine heliotherapy and of
the curative properties of the combination air-sea-sun. However,
there is also a marine thermal medicine, called thalassotherapy,
which has achieved appreciable results in the treatment of several illnesses among them rheumatism, healing of bone fractures,
and rhinitis. Drugs from the sea constitute a segment of the
pharmacopoeia. A symposium was organized in 1970 dealing exclusively with the ocean as a source of useful drugs. Many old medical treatments are based upon the use of marine plants. Thalassotherapy stations are not of recent vintage. There are several
such cure centers in France (Roscoff, Biarritz, Quiberon), in
Germany, and at Ostend, Belgium, and others.

Under given conditions, sea water can be administered to
patients orally, or by intravenous injections. Hot and cold
baths, often accompanied by high pressure water streams, have
been successfully used to combat obesity, neuritis and poly-
neuritis, lumbago, cellulitis, and nasal conditions. Thousands
of children were saved in pre-World War I days by sea water injections in specialized clinics at Paris, Brest, Reims, and Nancy,
in France. Nasal absorption of sea water has proven beneficial
in curing sinusitis. Yet, the medical use of sea water faded
away after the end of World War I. Just as there is a renewal of
interest in the Western World in Chinese acupuncture so there is
now a renewal of interest in thalassotherapy. At some marine
spas electrotherapy, thallassotherapy and acupuncture have been
combined in electro-thalasso-acupuncture with apparently remark-
able results.

1.4 DRUGS AND MEDICAL RESEARCH

Man has dipped into the oceanic medicine chest for centuries.
Ancient sea remedies were algae to help with stomach disorders,
and such organic products as agar, protamine sulfate, crude cho-
lesterol and cod liver oil. Even fossil material such as diatom
ooze and ichthyammol has been put to use. For many centuries
Japanese extracted iodine from algae and more recently the brown
algae's alginic acid is used to calculate tables indicating food

disintegration in the stomach, or spun into fibers for the manufacture of surgical swabs, dental plugs and gauze dressings. Red
algae provide agar which is used as a laxative and also has beneficial effects upon duodenal and peptic ulcers.

Especially over the last quarter of a century have the anti-
biotic properties of seaplants and animals been studied. A study
by C.P. Li of the National Institutes of Health (Bethesda, Md.)
shows that Paolin I inhibits bacteria, and Paolin II inhibits
viruses such as polio. Both are extracted from shellfish such as
oysters, clams and conches, and from squids, abalones and sea
snails. Dinoflagellates, causing the red tide phenomenon, also
contain several antibiotics. The staphylococcus organism which
resists penicillin is killed by sponge antibiotics and the green
alga's halosphaerin. Ectyonin, extracted from the redbeard sponge
is an effective microbe inhibitor. Eptatretin, a biochemical
found in the three-hearted hagfish, regularizes the heartbeat and
could be used for cardiac cases; it reverses arrhythmias and might
be considered as a chemical coordinator in the conduction processes of the myocardium in humans (D. Jensen of Scripps).

The electric eel possesses cholinesterase. Chemists can de-
rive from this enzyme pralidoxime chloride which acts as an anti-
dote to insecticide poisoning. Further study of this fish may
provide relief for neurological disorders. Amputees suffering
from destroyed nerve ends may be helped by holothurin from the
seacucumber because of its nerve blocking effect, while high
blood pressure is lowered by the stonefish's poison, and heart-
beats are slowed considerably by weeverfish venom. One of the
most virulent of sea denizens is a coelenterate, the seawasp.

Holothurin also shown anti-carcinogenic properties. Bonel-
linin is a growth inhibiting hormone secreted by the proboscis
of a sea worm; it could be useful in regulating cancer growths.
As for Neptunea antiqua, its gastropod poison is a potent muscle
contractant.

Tetradotoxin, from the puffer fish (fugu) family relieves
pain in terminal cancer; cephalotoxin, extract of octopus saliva,
inhibits blood coagulation and induces muscle paralysis, and this
animal's elecoisin controls high blood pressure and regularizes
heart beat after an attack. Conus geographus, a gastropod, provides a muscle relaxant, possibly an anti-convulsant, while
Conus magus produces paralysis thus helping contract diseased or
injured muscles; both toxins are reversible in effect. In a later paragraph we will speak about the spawn of the seaurchin.

Marine animals have also proven invaluable in medical research. Human nerve cells are very small and individual fibers
are hard to study, not so with the squids' basically similar

cells. The highly sophistacted octopus brain is quite appropriate
for brain ablation and drug reaction studies. The animal also ex-
hibits such emotions as anger and fear and can be used in some
basic learning research studies.

Fishes contract cancer, suffer from heart and liver diseases,
react to LSD, and are amenable to transplantion of organs.

Marine neurophysiology may lead to clues to help humans.

1.5 HARNESSING OCEAN ENERGIES

If there is a "water crisis", there is also an "energy crisis". Perhaps with the help of the ocean electrical black-outs of
the type experienced in New York and London can be avoided. Fossil energy production needed per person in 1800 was 300 calories;
by 1970, the requirements had reached 22,300 calories. The world
is facing an electrical energy crisis, stated recently Nobel
Prize winner Glenn T. Seaborg, former Chairman of the U. S. Atomic Energy Commission. Seaborg felt that the additional source of
energy should be the atom. Conservationists contend that nuclear
power plants are sources of thermal pollution whose effects are
not known. Another alternative often considered is the harnessing of the energies of the oceans, energy which is dissipated,
for instance, by tides, waves, and currents.

The current energy crises has, to use the words of a Miami-
based oceanographer, brought about a reversal of thinking at the
highest scientific levels. When one writer made a proposal in
1969 for an exploratory study of ocean energy tapping, it was
rejected. Today, these proposals are again under study. Research
divisions of power companies are now gathering information on
tidal power possibilities, and the U. S. Navy is being approached
for the funding of an ocean thermal power research project. The
U. S. Government's Energy Research and Development Administration
(ERDA) is also involved.

1.5.1 Tidal Power

The Ancient Greeks had already attempted to take advantage
of the Euripus tides, a channel separating Euboea and Boeotia.
Near Chalcis, water mills put the energy of currents to use,
while near Agostoli, on the isle of Cephalonia, energy to run
mills was obtained from the tides. Tide mills were in use in
England and Wales as early as the year 1000. Even in the United
States, in New England, and on Long Island, tide powered mills
were numerous. A major treatise on the subject of using the
force of the tides was published in France in the late 18th

century by Bernard Forest de Bélidor. But plans on a larger
scale were placed on the drawing boards after World War I. Possibilities of using the tides to produce energy have been examined in England, Wales, France, Australia, Argentina, the United
States, and elsewhere. Under President Franklin D. Roosevelt
work was actually started on a tidal power plant near Passamaquoddy, Maine, in 1935; this was halted when Congress failed to
appropriate funds. Under President John F. Kennedy, a similar
plan was revived and approved, but Congress failed again to appropriate funds.

Tidal energy is derived from the force inherent in the
earth's rotation and the gravitational pull of solar bodies. Only
France and the U.S.S.R. have actually built tidal power plants,
the former in the estuary of the Rance River, near St. Malo on
the coast of Brittany, and the later on Kisgalobskaia Bay near
the White Sea. (Fig. 10, 10a and 10b)

The French Rance River Tidal Power Plant which was completed
in 1966, is integrated in the French National Power Grid, and has
a net production of 537 million kilowatt hour in the basin to sea
direction and 71.5 MkwH in the sea to basin direction. With 64.5
MkwH used for pumping, this leaves a usable production of 544 MkwH.

The Soviet's plant on the Kisgalobskaia (Mezen Bay) in the
vicinity of Murmansk produces close to 400,000 kw.

Further excellent sites include the Severn River near Car-
diff in Wales, the Northwest Coast of Australia, and areas such
as Cabo Tres Puntas in Argentina. The Canadians have a permanent
Committee on Tidal Power. And the Passamaquoddy Bay in the northeastern United States is the region Roosevelt and Kennedy had in
mind. There are at least another 30 top locations in the world.
Of tidal power in the Northeast of this country, a National
Petroleum Council Report said singling out only a Bay of Fundy
plant: "the generation of 50 billion kwH annually ... would
correspond to 240,000 barrels of oil per day, or about 2.6 billion barrels over a 30-year period ... costs for equivalent fuel
oil ... would be in the range of $7 to $10 billion." While main-
taining that "potential for annual tidal energy is limited", it
continues by admitting that "nonetheless, the summation of tidal
energy over many years becomes significant", and recognizes that
"the ultimate potential in the Passamaquoddy-Fundy area is several times [that of the 50 billion kwH]." (Fig. 11)

A conservative estimate made by Walter Munk of Scripps
Oceanographic Institute   places the total global tidal energy
at 3×10^{12} watts, of which some 350 Twh would be harnessable.
The National Academy of Sciences report (1969 Resources and Man)
calculated 13,000 MW.

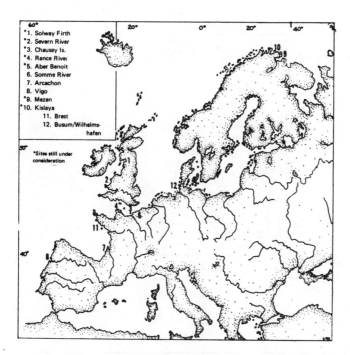

1. Solway Firth
2. Severn River
3. Chausey Is.
4. Rance River
5. Aber Benoit
6. Somme River
7. Arcachon
8. Vigo
9. Mezen
10. Kislaya
    11. Brest
    12. Busum/Wilhelms-
        hafen

*Sites still under
consideration

Sites of proposed tidal power plants

FIG.10

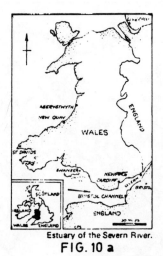

Estuary of the Severn River.
FIG. 10 a

The Severn Barrage.
FIG. 10 b

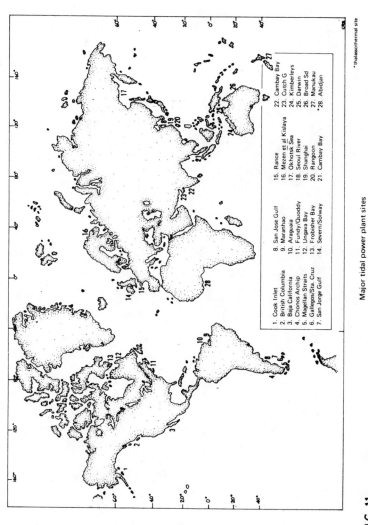

Major tidal power plant sites

FIG. 11

*thalassothermal site

1. Cook Inlet	15. Rance
2. British Columbia	16. Mezen et al Kislaya
3. Baja California	17. Okhotsk Sea
4. Chonos Archip	18. Seoul River
5. Magellan Straits	19. Shanghai
6. Gallegos/Sta. Cruz	20. Rangoon
7. San Jorge Gulf	21. Cambay Bay
8. San Jose Gulf	22. Cambay Bay
9. Maranhao	23. Cutch G
10. Araguaia	24. Kimberleys
11. Fundy/Quoddy	25. Darwin
12. Ungava Bay	26. Broad Sd
13. Frobisher Bay	27. Manukau
14. Severn/Solway	*28. Abidjan

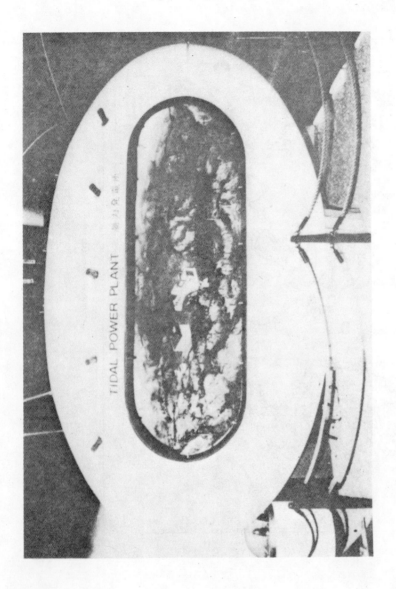

But whether such plants are really economic remains contro-
versial. While the Soviets are building a second plant in their
rapidly developing Europe Arctic zone, the Electricite de France,
ignoring the challenge given their report by leftist organizations, has again postposed a decision on construction of a new
plant near Mont St. Michel on Chausey island. Such plant could be
completed by 1982 and would save yearly eight million tons of oil.

The price tag of such plants has been the main deterrent to
their construction. It was the motivating factor in deciding not
to construct the Severn River plant in Great Britain; and, yet,
it is freely admitted today that, had the Severn plant been built
when first considered in 1933 or next in 1945, the operation would
have paid off handsomely within 10 years. In Australia, the possibilities are great and a plant built on the Kimberley coast in
western Australia would provide over 300,000 kw, or about 50 times
the present production of electricity of Australia. A negative
decision was taken because energy cannot be stored and there is
not yet a sufficient market in Southeast Asia to use that much
power . Meanwhile, the French plant has proved to be a boost to
the development of Brittany and a ·worthwhile addition to the
French national grid at peak demand times.

1.5.1.1 The Rance River Plant

In the world's first actually operating system, the difference of from 9 to 14 m between high and low tide produces over
544,000 kilowatts. Because of its reversible operation, power is
tapped in the Rance River plant from the waters as they rush upstream at high tide, and as the waters recede towards the sea.
Turbines thus generate power as the reservoirs empty and as they
fill.

The new station is being linked with France's national electric grid and the reservoirs' level can be raised by pumping. The
storage ability of the station is one of its most valuable assets.
The volume of water displaced reached 718 million m. The power
plant is 390 x 53 m. wide and, when in full operation, 24 generating units are put to work.

The principles are relatively simple. By cutting the estuary
with a dam, an upstream basin was created. Water rushed through
the opened sluices into the basin at incoming tides, and the flow
is stopped when the high tide is reached. Thereafter, the basin
is permitted to empty into the sea -- and energy is created. The
nonreversible blades also permit the production of electricity
as the tide flows in. Thus, around-the-clock production is obtained.

Started in 1963, and costing close to $100 million, the
Rance River project required the removal of 1.5 million m. of
water and the drying up of about 75 hectares of the estuary. The
dam accommodates a road which cuts 30 kilometers from the distance between St. Malo and Dinard, and eliminates a ferry. The
plant went into operation in November, 1966. (Fig. 12)

1.5.1.2 Canadian, U.S. and Soviet Plants

The Soviet Union has built an experimental plant on Kisgalobskaia Bay.The Soviet Union's small coastal bays hold a power
potential exceeding some 3.2 billion kilowatts. The plant was
completed in December, 1968, and has a capacity of 800 kilowatts.
Its engineer, Lev Bernstein, sees the success of tidal power stations in international cooperation. He said, "The tidal energy
of the English Channel, linked up with that of Mezen Bay and
coordinated, could supply peak power for all of Europe."
(Soviet Life, No. 160, Jan. 1970, P. 37)

The Soviets are now starting a commercial plant near Lum-
bovka on the Kola Gulf and on the Mezen Bay. Capacity of these
stations will be respectively 300,000 kw and 25,000,000 kw, with
tidal amplitudes in these gulfs of 7 and 9 meters respectively.
They are also searching for power to industrialize Eastern Siberia;
by harnessing tides in the Okhotsk Sea bays of Penzhina and Giz-
higa, whose amplitudes reach up to 13 meters, the Soviets could
generate 174 billion kilowatt hours.

Canada has a tidal power plant program under study for the
estuary of the Memramcook in the Bay of Fundy; and, on the Canad-
ian/U. S. border, sites have been under consideration (especially in Passamaquoddy Bay) for nearly 50 years.

The Passamaquoddy Bay has tide differences reaching 15m.
Major objection to the tidal energy station project has been -
like in Australia - the distances between the plant and the even-
tual consumer. Since transportation no longer is a problem, interest in the project has revived. If ever completed, it would
dwarf the Rance River project.

The multiple-basin plan at Passamaquoddy actually went into
construction in 1935, but work came to a halt when Congress
failed to appropriate money. Congressmen had the same attitude
in 1963 and 1964.

Among the main advantages of tidal power are its regularity
from year to year with less than 5% annual variation. Though
there are substantial disadvantages (see: Roger H. Charlier,
1970, "Harnessing the Energies of the Ocean", Marine Technology

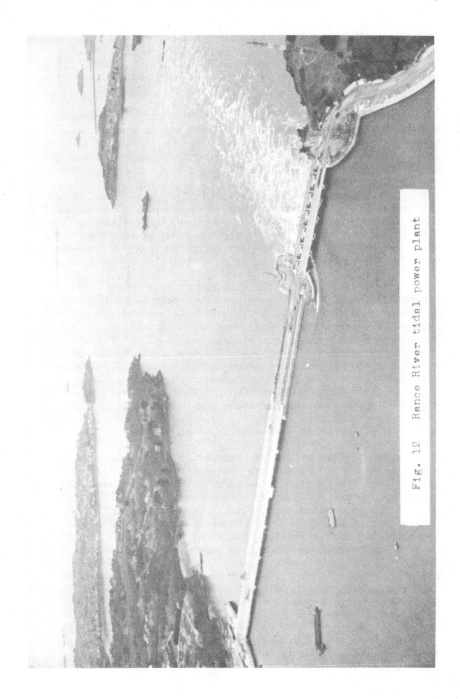

Fig. 12   Rance River tidal power plant

Society Journal, Vol. III, No. 3, pp. 13-32 and No. 4, pp. 59-81,
such plants permit simultaneous use of the dam for a road or rail-
road, provide estuary navigation improvements, production of cheap
electricity, a virtually inexhaustible supply of energy, and constitute encouragement for unexploited areas. And it is totally
pollution free.

A decade ago, nuclear power seemed to be less expensive,
though capital investments costs are of the same magnitude as
those for tidal energy plants. Yet, the new machinery now available has reduced drastically the competition between these two
sources of power.

R. G. Voysey sees the possiblity of an ideal "mix" - perhaps
a happy marriage - of nuclear and tidal schemes. The nuclear
plant could work continuously at full load, hence putting its capital investment to 100% use, while the tidal power plant could
store and deliver power "with an eye to the value of peak-demand
power".

Major competition to the tapping of ocean power still comes
from nuclear plants. Offshore nuclear centrals can be built on
land and towed into place, a procedure followed by the Soviets
with their experimental tidal power plant as well. The Netherlands have been working on a plan to build a huge artificial island in the North Sea which would house a nuclear power plant, a
garbage dump, and polluting heavy-industry plants. Tentatively
sited some 60 km. north of Rotterdam it could double as a ship
terminal for vessels reaching 400,000 tons. Oil refineries, iron
and steel works, aluminum shelters and ship repair facilities
could be accommodated. A similar plan exists in Belgium to build
an offshore complex at some distance from the coast.

Another factor militating against tidal power plants is the
strong emphasis placed by some economists on an efficient allocation of resources and their concern for making a choice between
competing demands for limited resources. Placing the major importance on discounted cash return they obscure the real issue behind a veil of possibly faulty numbers concerning interest rates,
secondary benefits, transfer functions, exhaustion or increase in
cost of alternative power sources and pollution abatement requirements. Suitable cost-benefit analyses have shown the economic
rationality of such projects, but in capitalistic countries doubt
is being raised, mostly by economists working for investor-owned
utilities, as to whether government participation in power development is in the best public interest.

While the time for expanded tidal energy harnessing may well
have come, ocean power is available from several other sources.

1.5.2 Thalassothermal Power

The ocean can also provide energy if we harnessed the power
dissipated by waves, currents, upwellings, thermic exchanges,
fresh- and- salt water contact, salinity exchanges, the marine
electrical energy from solar radiation (.04 watt for each cubic
meter of water), the heat flow transmitted by conduction through
the ocean bottom, the dissipation of kinetic energy caused by
surface agitation brought about by the wind.

Electromagnetic energy seems rather limited when compared to
the total energy available. Experimental plants using water temperature differentials to produce electricity have been constructed and operated. The late Georges Claude, the French engineer,
built such a plant actually using thermal pollution.

Thermal energy results from temperature differences between
two water supplies of unlimited discharge. Warm surface waters
in equatorial regions and in the tropics and deep cold waters
flowing from the polar regions come into contact in these areas.
An experimental plant was built by the Société d'Energie des Mers,
in Abidjan, Republic of Ivory Coast, but was abandoned after a
short period of time. Such plants, whose most ideal locations
are upwelling zones, near coasts, in tropical areas exposed to
winds of constant direction, nevertheless hold great promise.
Their production of electricity can be coupled with fresh water
production, air conditioning and extraction of sodium chloride,
sodium sulfate, chlorine and hydrochloric acid. Recently, Barjot, a physicist, discussed the possibility of building thalassothermal plants in polar areas where the difference of temperature between the ocean water and the overlying air layers reaches some 50°C. Neither the tidal power plants nor the ocean
thermal plants interfere with the environment. They are pollution free sources of energy.

Georges Claude made several thalassothermal attempts, the
first one using thermal pollution caused by manufacturing plants'
waste waters in the River Meuse near Ougrée, Belgium. He made
further attempts off the coast of Brazil as well as in Cuba. But
credit should also go to Charles Beau and N. Nizeri, both Public
Works engineers, who founded the Société d'Energie des Mers which
actually built the first full scale operating plant utilizing the
temperature differences between deep and surface ocean waters.
Actually, theirs was the culminating effort of a long series of
unsuccessful attempts stretching over the twenty-year span separat-
ating the two world wars.

The full scale attempt was initiated in 1942 by the French
Ministry of Colonies and the Centre National de la Recherche
Scientifique. After six years of study, construction of a plant

was finally decided. The technical aspects are abundantly described in a report published in the Proceedings of the Fourth
World Power Conference (1952). Basically, in this project, the
thermodynamic cycle consisted in evaporating, under vacuum, part
of the warm surface waters at 28°C (82°F), encountered in tropical areas. The steam is taken in by the condenser, itself cooled
by the colder (8°C - 46°F) deep waters, and on its way the steam
proceeds through a turbine that drives an electric generator.
At that time sea thermal energy compared favorably from an economic viewpoint with hydro-electric energy, the more so since the
energy could be directly used for an evaporating plant for chemical industries.

The site chosen was Abidjan in today's Republic of Ivory
Coast. Here fresh water was in greater need than electrical energy. A project combining both needs was ideal. The major problems arose with the immersion of the cold water adduction pipe;
repeated experiments with a large diameter duct all failed. When
Nizeri used articulated joints for the duct and anti-wave float-
ers to hold it up suspended on cables held by winches, the operation met with success. The plant nevertheless went out of business due to the fact that conventional power plants produced
cheaper electricity. The ducts suffered repeated ruptures and
the turbines required.large dimensions. It is noted, that Hydronautics' new system calls for no less than six 23 meters (78 feet)
turbines using temperature differences of 2.2C (36°F), while the
French engineers felt a difference of less than 15°C (59°F) should
not be considered. The Hydronautics' system is suitable for a
wider geographic range. Since Beau and Nizeri's scheme, the matter has been re-examined in France by Gougenheim and Romanovsky
(1957) and Daric (1957). Howe conducted studies dealing with
size and cost of turbines at the University of California at
Berkeley; combination solar-thermal energy schemes were proposed
by Barjot (1971), Rasson (1960) and Gomella (1966). Considerable
thought has been given to tapping ocean thermal differences as
well in the United States. W. Heronemus, at the University of
Massachusetts following the work of the Andersons, suggested
(1974) a scheme which envisions a series of such plants spaced in
an area 15 miles east-to-west by 550 miles south-to-north along
the western portion of the Gulf Stream. The electricity produced,
according to Heronemus, could be transmitted to virtually any
U. S. location at competitive cost. The plants would use a closed
Rankin cycle with ammonia or propane, for instance, as a cooling
intermediate fluid, which could function with a 17°C (62°F)
temperature difference. The power plant itself would be semi-
submerged and contain multiple units. J. H. Anderson, and his
son J. H. Anderson, Jr., founders of Sea Solar Power, Inc. pio-
neered many of the current ideas in this field of oceanic electrical production. Credit for proposals in thalassothermal

development goes also to A. Lavi and C. Zener. Zener, then Chief
Scientist at Westinghouse, studied the practical aspects of tapping ocean thermal power in 1965; Lavi and Zener are re-examining
the open cycle option while the Division of Solar Energy in the
U. S. Government's Energy Research and Development Administration
(ERDA) is forging ahead with the closed cycle system. Finally,
it should be pointed out that Lavi and Zener's views are parallel
with the Lockheed's study conclusions, though there are divergen-
cies where water pipe and heat exchanger designs are concerned.

In 1970, Gerard and Roels published in the Journal of the
Marine Technology Society a paper concerning ocean water as a
resource. In it they explored using upwellings in the Virgin
Islands as a cold water source to provide a temperature difference with the surface waters to produce electricity. The study
also involved fertilization of the sea. While production was
very successful, using the warm waters, ecologists object stren-
uously, warning of environmental destruction, multiplication of
algae, invasion of starfish, and possible ruin of the coral reefs.
Roels, however continued with his studies, in collaboration with
Othmer (Science, 12 October 1973). They feared that the various
designs involving vertical suction pipes suspended from vessels
or platforms, submerged power cables and fresh water lines carrying products to shore would augment the difficulties for controlled mariculture, which they propose to link with energy production. They also discussed in detail the engineering design made
by Alemco, Inc. for their particular site.

The idea of harnessing ocean thermal energy has been examined for nearly a century, but less than ten years ago the
interest was still mostly academic. Current and future energy
needs have instilled a keener interest in thalassothermal development.

The interest in such power plants has been recently rekind-
led. Much thought has been given to using an intermediate fluid.
The idea is not new since d'Arsonval suggested ammonia in 1881,
Campbell liquified gas in 1913; later came propane and R-12/31
(an Allied Chemical refrigerant). At least six designs have been
proposed. The United States pavilion featured at OCEANEXPO-75
the Ocean Thermal Conversion System, Delta-T. The success of
thalassothermal plants is no longer dependent on feasibility, but
on circumventing some remaining construction, operation and
possibly environmental problems. Engineering firms are also con-
vinced that ocean-thermal plants can compete, price-wise, with
conventional electricity producing plants. The United States
Government has put some 38 persons to work on thermal energy of
the sea projects and appropriated several million dollars for
such research.

1.5.3 Solar Power

Taken into consideration that bottom waters remains close to
3.9°C, whether in tropical or polar areas, while air temperatures
in polar zones stay near -45°C.   Physicist Barjot suggested
that polar waters could be used as a source of warm fluid, while
the air could provide the cold fluid. This would yield a temperature amplitude of about 50°C, or double the difference available
in the watery domain. Furthermore no long water ducts would be
necessary. The Barjot system would use butane, a gas that does
not mix with water and liquifies at -10°C. Thalassothermal plants
based on the Barjot principle could be built in Scandinavia,
Siberia and polar areas in general. No environmental or ecological disturbances would result.

Solar energy could be used to overheat ocean surface water
covered with a thin oil film in shallow basins. No evaporation
would occur; adduction conduits would not have to reach into considerable depths (300 to 400m). Masson studied the latter possibilities near Dakar, Senegal (Masson, H., 1955, L'utilisation de
l'énergie solaire dans les regions arides: Industries et Travaux
d'Outre-Mer III, 17, 226-232); ibid., 1960, L'énergie solaire
et ses applications: Annales des Mines IV, 4, 163) Hirschmann
drew blueprints for a plant that would utilize near surface water
even as warm as 20°C and ocean water heated by solar energy to
about 70°C in a basin with an electric turbine placed between
evaporator and condenser (Gomella, C., 1966, La soif du monde et
le dessalement des eaux: Paris, Colin). Hirschmann's scheme is
coupled with distilled fresh water production (Table 2).

Elisabeth Borgese in her book on The Drama of the Oceans
(Abrams) quotes J. Hilbert Anderson and James H. Anderson as
estimating construction costs for a thalassothermal energy plant
at $200 per kilowatt versus the needed $700 per kilowatt for a
nuclear plant. And she adds: "The more pessimistic forecast of
$1100 per kilowatt would still make sea thermal power economical-
ly competitive with other power producing systems by the late
1980s". Anderson and Anderson estimated that the Florida Straits,
mentioned as an important source of ocean current power, could
provide 20 million megawatts by tapping temperature differences
of potential. Problems occurring with thalassothermal schemes
include corrosion, organisms fouling, dissolved gases removal,
energy consumed for pumping purposes and maintenance costs of
bulky turbines. Partial solutions have however been found, such
as using low boiling propane as an intermediate fluid which would
permit a five-fold turbine diameter reduction, from approximately
7 to merely 1 1/2 meters, and submerging the condenser and the
boiler at depths of approximately 50 and 85 meters respectively
so as to balance inside and outside pressures and thus make possible the use of thinner walls.

Table 2. Two-Group Plant Producing 7000 kw Net

Cold water discharge (mc/sec)	10
Warm water discharge (30°C) (m3/sec)	30
Temperature drop at evaporator (C)	3
Evaporation temperature (C)	27
Mean temperature at condenser (C)	16
Mean difference (C)	11

ELECTRICAL POWER PRODUCTION (kW2)

Total hourly energy produced	11,000
Net yearly energy produced	56,000,000

FRESH WATER PRODUCTION (m3)

Hourly fresh water yield	600
Daily fresh water yield	14,000
Yearly fresh water yield	4,800,000

COSTS (US $)	1957	1969	1974
Electrical power plant	7,860,000	14,000,000	20,930,000
Additional for surface condenser needed for fresh water production	546,000	1,000,000	1,500,000

1.5.4 Ocean Currents Power

Ocean currents power is considerable but diffuse due to low
velocity and has heretofore generated little interest. An additional factor causing problems is that currents vary their precise position. A cluster of turbines suggested to tap currents
energy in the Florida Straits by von Arx, Stewart and Apel could
furnish about one million kilowatts on a year-round basis, or as
much as two very large nuclear plants. The turbines would, however, be bulky, so maintenance would be costly. The idea, is
sound and such plants resemble hydroelectric plants built on
rivers. The current-power station could tap the energy of the
narrow and intense flow of western boundary currents of major
currents gyres. The "water low velocity energy converter",
Gerald Steelman's brainchild, would place an electric generator
on an offshore anchored ship. The current drives a wheel below
the vessel and the cable is equipped with umbrella-like devices
which open when facing the current and collapse otherwise. Taking into account both principles of moving waters as in river
hydroelectric plants and of contacts of water masses of different nature as in thalassothermal plants, is an idea proposed by
Scripps Oceanographic Institution's J. D. Isaacs. The sun causes
sea water evaporation, thus actually separating fresh water from
sea water. Water vapor eventually condenses and falls upon the
land. Part of it finds its way to rivers and when river water
meets the sea, the solar energy used for the evaporation becomes
available anew. The freshwater/salt water contact creates a difference of potential whose energy equals that of a 240 meter
waterfall. "Salinity power" is also used in electric batteries.

1.5.5 Wave Power

Wave power is provided by the onslaught of a breaking wave
which can be captured in a reservoir, accessible by way of a converging ramp, and connected with a return channel at the exit of
a low pressure turbine. The power can also be generated by means
of devices set directly in motion by the wave itself. Experiments with wave power have been conducted in Pacifica, California,
in Chicago, in Boston, in Biarritz, and in Japan.

Ocean waves represent a significant source of energy. They
are a form of solar energy created through the interaction of
winds with the ocean surface. The power available in such waves,
although relatively diffuse, is impressive when one considers
that there is more power represented in the potential energy of
a heaving ship than is present in the thrust of its engines. Using the Atlas for Mariners (US Navy Hydrographic Office 1959),
Walter Schmitt calculated that the available power from ocean
wind waves is 2.7×10^{12} watts summed over all of the oceans.

Isaacs and Seymour (1973) calculated a figure of 2.5 x 10^{12} watts
by summing an average surf condition over all of the world's
coastlines. This amount is roughly 1/10 of projections of the
global power requirements for the year 2000 A.D.

Although no one would propose covering the oceans with devices converting this power to a more useful form, the very nature of wind wave energy requires a great number of small devices
for its extraction. Furthermore, wind waves have a unique property that makes more energy available as energy is extracted. This
is due to the relative inefficiency at which energy is transferred from the wind to the sea at highly developed sea states.

In 1909 Alva L. Reynolds made a first attempt at converting
wave energy into electrical power. Several patents were filed
during the early nineteen hundreds. Reynolds built a wharf and
suspended panels beneath it. The force of the waves was transmitted to a wheel attached to an electric generator. Storage
tanks fed a water pump by trapping waves.

Though functional, Reynold's machine was not economically
practical because not sufficient power was captured. A wave
motor was tested off Pacifica near San Francisco three years ago.

Waves, if sufficiently regular and powerful, could provide
an acceptable source of energy. Waves commonly are small in
height and return every few seconds, but severe storms, which
cannot be either avoided or predicted, require special constructions which are both very expensive and of little return.

Converging wave channels, supplying a basin constituting the
forebay for a conventional low head power station, seem to provide the highest output of any scheme proposed to recover wave
energy. Although technically feasible, such a power station
appears uneconomical.

Ten years ago, scientific reports were made on the available
energy of generators designed along the same lines as conventional aerogenerators. Wave power has been recently brought
again to the foreground by a Boston, Massachusetts firm ("A New
Concept in Tapping Wave Energy", Ocean Industry, Vol. 4, No. 3,
pp. 62-63, 1970).

The power of ocean waves is considerable. If it can be put
to work to produce electricity, waves could be tapped to produce
kilowatts. Lybrand Smith, of the U. S. Navy Bureau of Engineering, figured that 33 CV per meter (10 hp per foot) of sea wave
front are furnished by a wave less than 2 meters high. This
represents more than 31,500 CV/km (50,000 hp per mile). And,
in 6 meter deep water, a 3 meter wave has 115.5 CV per meter

(35 hp per foot) of sea wave, equivalent to 116,550 CV/km
(185,000 hp per mile).

In 9 meters deep water, a 3.6 meters high wave would provide
202.26 CV per meter (61.29 hp per foot) of wave front, or 204,120
CV/km (324,000 hp per mile). This represents roughly 150,000 kw
per kilometer (242,000 kw per mile).

Exceptional tidal amplitude is not very common throughout
the world but waves are widely available. Numerous large cities
are in coastal areas, hence wave energy is present for utilization close to consumption centers. Waves occur in inland bodies
of water as well, but their power is 2 1/2% less than that of sea
water waves.

A Boston firm, Power Systems Company, patented a wavespower
system and conducted successful small scale tests. The basic
principle involves a concrete trough parallel to the shoreline
in which a pliable strip filled with a hydraulic fluid is secured.
Installed at a depth of 7 meters, so that wave shape plays no
significant role, the strip breaks the waves, undergoes the hy-
drostatic pressure of the water mass above, and transmits this
pressure to an hydraulic accumulator through the hydraulic fluid.

The accumulator stores pressure until a specific magnitude
is attained, then delivers it to a fluid motor, connected to a
dynamo which generates electricity. Tide level does not influence power production.

The designers point to a long list of advantages of wave
produced power: 1) it is pollution free; 2) it is widely available; 3) since the original power is free, low cost operation is
provided; 4) additional units, furnish additional power; 5) sites
can be selected on unused shore land; 6) installations would constitute protective structures for coasts and harbors; 7) the
power generator is much more efficient than those using fossil
fuels; 8) present power plants, located mostly inland, will, in
the near future, be unable to satisfy consumers' demands, hence
wave power stations, like tidal-power stations, would be support-
ive; 9) output is unaffected by weather or climate; 10) because
climatic and meteorological data is readily available, size and
power of waves can be predicted; 11) the power units can be coupled to desalination plants; 12) wave power stations will not substantially disturb natural environment nor local ecology.

1.5.6 Other Types of Ocean Energy

Wind itself may be tapped as an indirect form of solar radiation. Two deterrents are the need to anchor windmills at sea

and the wind force variability. Favorable geographical sites are few.

An ocean-powered plant could provide the energy necessary to extract hydrogen from sea water. Easily stored, transported and available everywhere, hydrogen could be the basis for an entirely new economy. Except for the floating platforms, the environmental impact would be nil.

Deuterium can be used for nuclear energy production. Its presence in ocean waters may well be an additional incentive to locate nuclear power plants offshore.

The geothermal energy associated with rift zones and volcanic activity has generally been considered as of little prospective value in areas far from land because of the difficulty of utilizing it. However, in coastal and insular situations, geothermal power could be tapped to produce electric power; such plants could possibly be coupled with desalination operations; harnessable geothermal energy is furthermore likely to be found in zones where petroleum is also present. It is pollution-free. Waters in the northern Gulf of Mexico possess both thermal energy and methane.

2. BIOLOGICAL RESOURCES

Putting water itself and the production of energy aside, ocean resources can be classified into three categories: biological, chemical, and geological. The idea of resource can of course be extended to include the ocean bottom and its edges, offering sites for cables, pipelines, harbors and resorts, and the water itself provides maritime routes. During the Oceanographic Congress held in Moscow in 1966, some oceanographers insisted on the inexhaustible quality of ocean bioresources insisting even that the then current catch of 60 to 80 million tons of animal products could be increased to 200 million. Yet, it appears that the productivity of the Atlantic and Pacific oceans has reached its peak. There is a possibility of increasing the catch in the Indian Ocean and Antarctic Ocean.

Ocean fish resources are not limitless, or to use a dry humor quotation: "The ocean is not a great bouillabaise with infinite goodies to be caught. It is not a great cornucopia". But, if the ocean is not the panacea that alone can solve the hunger problem in the world, it can resolve, in part, the protein deficiency of some diets and certainly can provide more than the present 1% which it now contributes to worldwide nutrition.

While fish catches increase, due to the availability of

better vessels and more sophisticated devices, we are neglecting
long range problems.   Instead we should encourage diversification
of the ocean product: consuming other types of fish now considered "trash-fish" and marketing more shellfish. This cannot be attained quickly. To increase new ocean products consumption,
education concerning unutilized species is necessary. The more
so because more abundant living resources, at present little used,
such as krill and pelagic squids, thriving largely in areas <u>beyond</u>
the proposed economic zone (See V) will become accessible through
new technology. Appropriate international arrangements are needed
to insure an equitable future utilization and husbanding of these
"unconventional" resources far offshore.

The use of the biological resources of the ocean is a function of its productivity, of its accessibility (because refrigeration is far from widespread), and of the cultural heritage.
The plant life of the sea surpasses that of the land, but algae
are the principal constituent. Mexicans harvest kelp, Japanese
deepfry them, Bretons rake them from the beaches for fertilizer.

Of the marine plants Irish moss (<u>Chondrus crispus</u>) and nori
(<u>Porphyra</u>) are in demand. In the Orient they are used in soups,
salads, frys, etc. The main use of red and brown algae is in
the manufacture of agars, algin derivatives and carrageenan
(Table 4C). Agar and carrageen are extracted from red algae
while algin is a derivative of brown kelp (Macrocystis).

It is thus probably toward the source of proteins rather than
toward vegetal matter that our attention should be turned. In-
fant deaths are due, in major percentages, to the lack of proteins
in their diets, even though 3.5 to 7 kilograms of proteins per
year would suffice. An investment of about $3.00 per year would
provide a sufficient quantity of proteins per child, in the shape
of food protein concentrate (FPC). A bit less than half the total
catch of American fishermen is used to make fodder for cattle,
and developing countries process their seafood products in order
to sell them as pet food in foreign countries, an incredible waste
as far as human consumption is concerned since one ton of fish
represents 115 kilograms of protein.

2.1 CATCHING FISH

Archaeological and paleontology show that man has existed many
thousands of years, and it is apparent that fish has long held a
significant place in human diet. Early man fished with his hands
and feet, scooping and kicking fishes out of shallow pools where
they had been stranded by drought or receding floods or tides.
Fishes such as salmon spawn in small depressions in stream beds
called redds. From shallow redds, fishes were easily captured by

splashing and kicking.

Long pointed sticks and hunting spears proved helpful in
impaling fishes in pools that were too deep for hand fishing.
Multipronged instruments such as bidents or tridents proved even
more effective. The barbed sternum of a bird or a reindeer horn
fastened to a long pole formed a spear or a harpoon, and crude
spears with rude flint heads are evident from Stone Age relics.

Early man was not always successful in catching fishes with
his primitive spears. A quick stroke of its fins would send the
fish out of the spear's range. Provoked by disappointment he devised other methods. Keen observation showed that certain kinds
of fishes fed voraciously on smaller species. They also fed
greedily on insects or worms that fell into their waters. From
these observations, it is likely that early man created the first
fishhook, which was not a hook but, in reality, a gorge, a piece
of flint or bone that the fish swallowed with the bait and was
not able to spit out.

2.1.1   Development of the Fishhook

Gorges of various shapes have been found among the primitive
relics from early cave dwellings around the world. They usually
were a narrow strip of stone or a chip of flint with a groove in
the middle through which a string or line could be fastened. The
gorge would be buried in bait and swallowed end first by the fish.
Then, the fisherman would tighten the line so the gorge would become wedged crosswise in the fish's stomach or gullet and the
fish could be pulled ashore. A similar device was still in use
in the early twentieth century in England and France to catch
eels. In a fishing method called sniggling, a needle buried in
an earthworm serves a role similar to the prehistoric gorge.

It is easy to infer the development of the early fishhook
from the primitive gorge, as it became more and more curved.
Materials such as human and animal bone, slate, flint, shell,
cactus thorns, spruce wood, willow, and quills of feathers fashioned with a barb for impaling the fish are actually examples
of primitive fishhooks.

Curious fishhooks of ancient civilizations include the 1 5/8-
inch long chitinous, spurred hind leg of an insect, Eurycantha
lato, of New Guinea. The upper mandible of an eagle notched
down the base was recorded as a fishhook from prehistoric lacus-
trine deposits. North American Indians used bird claws and the
bones of cod and haddock as fishhooks.

2.1.2   Evolution of Fishing Methods

It seems likely that the first fishing lines were dried
plant tendrils, dried seaweeds, silk, and spider webs. Ancient
Greeks used the hair of horses' tails, and old Egyptian papyri
tell of the use of human hair for fishing line.

Chinese literature from Shik Ching, or Book of Odes, describes silk threads formed into fishing lines from the eleventh
to the seventh century B.C. Shik Ching states, "With your long
and tapering bamboo rods you angle in the 'Ch'i'" (a river in
Honan). These bamboo fishing poles were thought to be about 6
feet long. The use of fishing rods was found to be a great aid
in getting the line and hook out a distance from shore. The rod
also increased the leverage in setting the hook into the mouth of
the fish and speeded up the process of landing the fish.

In addition, early man captured fishes by luring them into
traps such as brush weirs, fashioned out of twigs and sticks, and
cages, made of woven vines and branches. With invention of the
net, fishes were either scooped, entangled, surrounded, or drag-
ged to the shore. Evidence of fishing nets goes back to the
Stone-Age settlements of Denmark and Sweden. R. Monro in The
Lake Dwellings of Europe mentions fishing nets from Robenhausen
and Vinetz - both belonging to the late Neolithic Age.

Netting was widely practiced in the Nile by ancient Egyptians.
The net appears in early hieroglyphs from the First Dynasty. Practically every type of net known to the ancient world was employed
by the Egyptians. Woven fishing devices depicted in Egyptian
writings and drawings include cast nets, stake nets, seines, and
wicker fish traps.

According to William Radcliffe in Fishing from Earliest Times,
spear, net, line, and fishing rod flourished simultaneously in
Egypt in the Twelfth Dynasty, about 3500 B.C. The Assyrian repre-
sentation of Gilgamesh carrying fish dates from at least 2800 B.C.

Among the most incredible nets are those used by the primitive people of New Guinea in the South Pacific. In Two Years
Among the New Guinea Cannibals, A. E. Pratt describes fishnets
made from spider webs. The webs spun by spiders in the Papuan
forests are 6 feet in diameter, with meshes varying from 1 inch
at the outside to about 1/8 inch at the center. The skill of the
spiders in web making has been put to use by the aborigines into
making fishnets. The natives take long bamboo poles and bend them
over to form a 6-foot oval loop. The large bamboo loops are
placed upright in the region where the spiders are abundant. In
a short time, the spiders weave their web within the perimeter of
the bamboo. The spider's web in the frame is water resistant and,

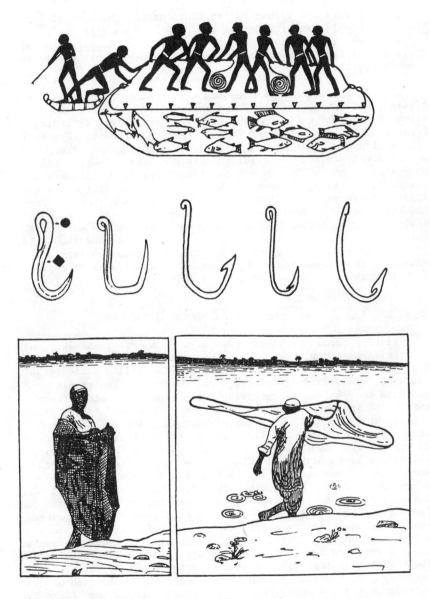

Fig. 12 A   Predecessors of today's fishing gear.

when dipped in water, will hold fish to a pound in weight.

Fish spears and fishhooks are mentioned in the book of Job
XLI. The New Testament mentions the fish nets used in the Sea of
Galilee (Lake Tiberias or Lake of Gennesaret). The Talmud says,
" ... in the Sea of Tiberias fishing with hook and net was everywhere allowed".

Today, the methods of catching fish - scooping, hooking, en-
closing, and entangling - are not vastly different from those employed by ancient civilizations. They differ mainly in their
greater size, efficiency and technological refinement. Synthetic
fibers of nylon, dacron, and polypropylene give added strength
and longevity to today's fishing lines and nets. The conical-
shaped trawl net pulled by motor-driven draggers or trawlers accounts for 80 to 90 percent of the world demersal fish catch taken
today. Purse seines, which encircle the schools of pelagic shoal-
ing fishes such as menhaden, pilchard, anchovies, herring, mackerel, and tuna, account for the major poundage of the world fish
catch. But, using handlines, hooks, and poles, around the world,
in the course of a year, are over a billion anglers - from pre-
schoolers to octogenarians and older - who still seek to capture
fish in much the same way as did primitive man.

2.1.3   *Fisheries and the Fishing Industry*

Fishing remains even in these days of modern technology, an
enterprise based on nets, hooks, human labor, and a good deal of
luck, yet the use of biomarine resources is still a sociological,
cultural, and political matter. Areas which are rich in fish
products are jealously protected by the riparian states. Peru
strives to increase its anchovy catch from one-half to ten billion tons; the Soviet Union has pressed into service factory
ships which accompany the fishermen's craft. Iceland has extended its territorial limits to keep other nations from exhausting
the waters surrounding the island. The United States, once the
second most important fishing nation, has dropped to sixth place,
behind Peru, Japan, China, the U.S.S.R., and Norway. The fishing
industry of the Soviet Union has made amazing progress in the past
few years. Its flotillas of over 100 fishing vessels are exceed-
ingly efficient; it has built new fishing harbors, new factory
ships, new spotting units, new harbor-to-fleet commuting vessels,
and has provided developing nations with ultra-modern trawlers.
The Soviets consider fishing as a very important source of hard
currency and, for that reason have reduced the production of
salted, smoked and dried fish, and increased the sales of fresh,
frozen, and canned products. So far, however, they have not
signed conservation agreements.

Soviet techniques have made considerable progress. In the
Caspian Sea fishermen use spotlights to find the fish, then use
suction pumps to bring them aboard, while the sonar is used in lo-
cating currents, upwellings, fish schools, and in checking bottom
features. One submersible is used exclusively for fisheries research. Both Japan and France have also improved their technological equipment. France has put to sea large vessels for tuna
fishing; these vessels carry a helicopter whose task is to spot
schools of fish and to direct the fishing operations.

Fish concentrate primarily in the shallow waters above continental shelves and near banks, near shore broad submarine elevations, and much of the fishing occurs in waters of depths not exceeding 400 meters where nutritive material is provided by benthic alga, land waste and plankton. Of world-wide economic importance are the non-tropical Northern Hemisphere fishing grounds,
although tropical fish catches are of significance on a regional
basis. Geographers recognize four major fishing areas: (1) the
coastal waters of Japan, Sakhalin and Eastern Siberia; (2) offshore New England, the maritime provinces of Canada, and Newfoundland; (3) Northwestern Europe including Iceland; and (4) the Pacific waters off Alaska, Canada and the American Northwest.
(Fig. 13)

In the late 1940's Japan alone caught 25% of the world's total
fish catch; this declined to 17% in the 1950's, and Peru overtook
her in the 1960's. The United States and Great Britain, once
Japan's closest rivals with about 7 1/2% of the world fish catch,
no longer hold this position. Among the commercially most prized
fish are herring, cod, tuna, bonito, sardines, mackerel and anchovies. (Fig. 14) The Japanese also use so-called "trash" fish
as fertilizer. The world's greatest cod fisheries are the North
Atlantic Banks stretching from Nantucket, Massachusetts, to the
eastern coast of Newfoundland. Here mackerel, herring, haddock
and halibut are also caught. While herring and mackerel are
caught by drift nets and lines (pelagic fishing), the others which
are deeper water swimmers (60 meters at least) feeding on bottom
invertebrates, require trawl lines, trawl nets, or hand lines
operated from the deck (demersal fishing).

The largest producer of canned salmon is the northwest coast
of the United States where each spring and summer millions of
adult salmon come to spawn in the rivers flowing into the Pacific
from the Bering Sea to Northern California. Only drastic conservation measures, enforced since 1920, saved the salmon fisheries
from extinction. Sturgeon fished for its meat and roe, with
stocks greatly depleted, is found from Scandinavia to the Mediterranean Basin, and from the St. Lawrence to the Gulf of Mexico.

The stormy waters of the eastern North Atlantic are plied

Fishing grounds
• Fish catches

FIG. 13

FIG. 13B. WORLD CATCH OF SEAFOOD
           BY LEADING COUNTRIES, 1956–1966

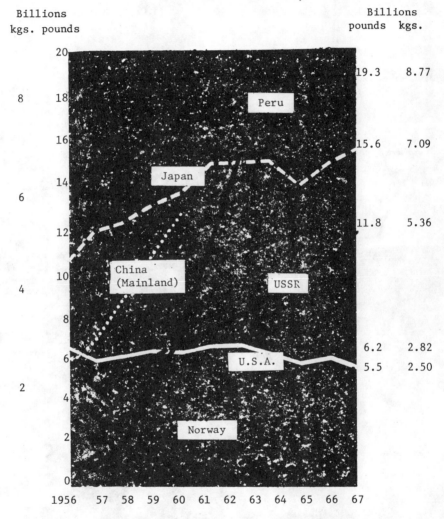

Note: Live Weight Basis
Source: Yearbook of Fishery Statistics – Food and
          Agriculture Organization of the United Nations
          Bureau of Commercial Fisheries Misc.

Fig. 14 American Tuna Fishermen

by British, Norwegian, Dutch, French, Icelandic, German, Portuguese, Belgian, Faroers and Soviet fishing vessels. (Fig. 15)
Though risky, this trade is the livelihood of a quarter million
fishermen. This year round activity, most intense in the spring,
bringing in cod, herring and mackerel, makes northwest Europe the
greatest fish exporting region of the world. The so-called cod
war with Iceland against other nations, particularly Great Britain, resulted from Iceland's concern with stock depletion.

Some tropical waters are rich fishing grounds where upwellings bring nutrient-loaded cold deep waters to the surface; here
production occasionally exceeds that of the major northern fishing grounds. Such areas are found off the West African coast,
near the Somali coast, in the Humboldt current off Peru. Fertile
tropical waters are especially fished for tuna in the open sea
and in coastal waters. Modern technology has also made it possible to fish in the high North Atlantic latitudes and the Arctic
(Tables 4 and 4A-D). Famed upwelling areas include the Peruvian,
Californian and Benguela upwellings.

Fish is eaten fresh, or it is frozen, salted, smoked, dried
or canned. It is also used to produce oil, meal, glue and fertilizer. (Fig. 16) Salmon, sardines and anchovies are the most
common canned fish; anchovies, menhaden and pilchards provide oil
and meal.

In 1975, a new firm located in New England, "Protein of the
Sea", had placed on the market minced salt fish selling for half
the price of the traditional dried salt cod, and by enforcing the
old adage "Waste not, want not" will reduce the pressure on the
demand for this food shelf staple. The product was placed on the
market after a new salt-curing method was devised which reduces
the time element from days to less than two hours; it needs no
refrigeration. Since the method is applicable to under-utilized
species, it will bring relief to an increasing demand from the
United States market and South America where Brazil alone buys
about 53,000 pounds a year of dried salted cod.

Freshwater fish will not be discussed here, though we mentioned salmon which is anadromous. Shellfish are exploited in
many areas of the world (shrimp, lobsters, oysters, mussels) including Japan, the Gulf of Mexico, the Chesapeake Bay (USA), the
Mid-Atlantic United States, Denmark, The Netherlands, France and
Spain.

Table 3 provides a general idea of the average catches of
leading fishing nations in the 1960's in millions of metric tons
A rough estimate per region follows.

Table 3. World Average Fish Catches Per Region

OCEAN/SEA	REGION		MILLION METRIC TONS
	NORTH		12
	CENTRAL	– WEST	24
PACIFIC		EAST	2
	SOUTH	– WEST	1
		EAST	1
	NORTH	– WEST	10
		EAST	31
	CENTRAL	– WEST	4
		EAST	2
ATLANTIC	SOUTH	– WEST	1
		EAST	4
	MEDITERRANEAN BASIN		3
INDIAN	WEST		3
	EAST		2

Fig. 15 Dutch Fishing Fleet

Fig. 16 Crated fish at Urk, The Netherlands

Table 4. Fish Catches in Metric Tons for Leading Types and Principal Fishing Nations

COUNTRY	SARDINE HERRING ANCHOVY	COD HAKE HADDOCK	MULLET JACK SEA BASS	TUNA BONITO MACKEREL	SALMON SMELT TROUT	FLOUNDER HALIBUT SOLE
Peru	7259	–	28	79	–	0.2
Japan	494	781	1455	1397	185	216
USSR	1347	1381	881	38	153	159
USA	841	164	258	160	159	106
Norway	1039	648	24	161	219	24
India	359	5.5	207	98	–	10
Spain	250	475	205	88	2	14
Canada	385	414	64	18	75	127
Iceland	763	730	38	–	50	10
Great Britain	161	702	33	3	3	64
Denmark	350	210	140	7	14	61
Chile	481	106	26	25	–	0.3
South Africa	413	87	67	54	–	2
Philippines	86	–	426	37	–	5
Mexico	31	–	32	18	–	0.3
Greece	31	4	21	6	–	2
Australia	0.7	–	24	11	–	0.1
Ghana	7.8	–	29	11	–	2
TOTAL (Above)	14,289.5	5297.5	3958	2211	860	802.9
OTHERS (Approx.)	4,571	1502	1750	618	380	339
Total (Approx.)	18,860	6800	5708	2829	1240	1142
PERCENTAGE	38	13.7	11.5	5.7	2.5	2.3

COUNTRY	SHARK RAY	MOLLUSKS	CRUSTACEAN OTHER INVERTEBRATES	OTHERS	TOTAL	%
Peru	8	5	0.6	5	7382	15
Japan	69	1138	134	981	6847	13
USSR	4	?	49	?	4309	9
USA	7	647	283	4	2634	5
Norway	32	11	14	85	2307	3
India	11	0.3	80	?	814	2
Spain	31	110	18	?	1330	2
Canada	0.4	89	21	31	1224	2
Iceland	0.4	0.2	5	3	1199	2
Great Britain	25	15	13	28	1046	2
Denmark	1.5	18	7	28	837	1
Chile	0.4	37	25	7	708	1
South Africa	0.5	2	8	31	684	1
Philippines	?	10	23	?	606	1
Mexico	5	41	62	82	252	0.4
Greece	4	5	2	36	112	0.2
Australia	5	17	18	8	29	0.1
Ghana	5	4	–	12	70	0.1
TOTAL (Above)	275.8	2149.5	762.6	1341	32,440	60
OTHERS (Approx.)	319	679	379	7146	17,193	40
Total (Approx.)	595	2830	1142	8487	49,633	150
PERCENTAGE	1.2	5.7	2.3	17.1	100	

Notes: Amounts, column-and row-totals rounded off and approximate.
Total average annual catch marine products 49,632,900 metric tons. (In the fifties catch was 23,624,200).
Total average annual catch including freshwater was 51,340,000 metric tons.
(After Oxford Economic Atlas)

Table 4A

World Commercial Catch of Fish, Crustaceans, Mollusks, and Other Aquatic
Plants and Animals (except whales and seals), By Species Groups, 1973–76

Species Group	1973	1974	1975	1976
	Thousand metric tons			
	Live weight			
Herring, sardines, anchovies, et al. . . . . ·	11,314	13,888	13,618	15,089
Cods, hakes, haddocks, et al.. .	11,970	12,699	11,882	12,116
Freshwater fishes.	9,293	9,244	9,599	9,532
Miscellaneous marine and diadromous fishes	8,676	8,382	8,021	8,445
Jacks, mullets, sauries, et al. .	5,740	5,454	5,935	7,389
Redfish, basses, congers, et al..	4,320	4,865	5,071	4,950
Mollusks.	3,459	3,424	3,779	3,917
Mackerels, snoeks, cutlass fishes, et al.	3,418	3,611	3,590	3,340
Tunas, bonitos, billfishes., et al.	1,999	2,125	1,976	2,209
Crustaceans	1,932	2,009	1,979	2,080
Miscellaneous aquatic plants and animals.	1,461	1,559	1,336	1,396
Flounders, halibuts, sole,et al..	1,248	1,175	1,143	1,123
Shads, milkfishes, et al. . . .	759	743	755	697
Salmon, trouts, smelts, et al. .	554	500	552	555
Sharks, rays, chimaeras, et. al..	589	544	571	533
River eels	52	55	58	64
Sturgeons, paddlefishes, et al.	24	25	27	31
TOTAL (1)	66,808	70,300	69,893	73,467

(1) Figures may not add to total because of rounding.
     Source: Food and Agriculture Organization of the United Nations (FAO).
              Yearbook of Fishery Statistics, 1976, Vol. 42.

Table 4B. Catch in Millions of Tons of Non-Fish Animal
Products in Order of Importance

TYPE	10^6 TONS	% OF TOTAL
Clams, cockles, scallops	1.000	19.43
Shrimp & prawns	0.930	18.03
Squids, octopus, cuttlefish	0.880	18.00
Oysters	0.730	15.27
Marine mollusks, exclusive of those listed above	0.690	13.88
Crabs	0.400	8.32
Marine crustaceans, exclusive of shrimp, prawns and crabs	0.290	5.55
Miscellaneous marine animals (e.g. worms, sea urchins, whales, etc.)	0.080	1.52
Total so-called "shellfish" and "other marine aquatic" animals	5.000	100.00
Share of this total in total marine catch		8.08

Table 4C. Harvest in Millions of Tons of Marine Plants

TYPE	10^6 TONS	% OF TOTAL
Red algae (Rhodophyta)	0.370	41.67
Brown algae (Phaeophyta)	0.330	41.66
Other marine plants	0.170	16.67
Total marine plants harvest	0.870	100.00
Share of marine plants in total marine bio-catch		1.39

Table 4D.   Total Marine Bioresources Gathered in 1970

TYPE	AMOUNT IN 10^6 TONS	% OF TOTAL
Marine and diadromous fishes	56.420	90.53
Mollusks	3.300	5.33
Crustaceans	1.620	2.62
Total "shellfish"	4.920	7.95
Other marine animals	0.080	0.13
Marine plants	0.870	1.39
TOTALS	62.233	100.00

There is no doubt that the biological resources of the ocean constitute a precious contribution to the feeding of a world in which hunger is an endemic condition for large segments of humanity. It is equally true that more nations should turn to the sea in their efforts to solve their food problems, but it is of paramount importance that no nation overlook the threat of overfishing which would rob the ocean of its regenerative capability. A primary step is to acquire a better knowledge of the capital which the wealth of the seas represents so that only a reasonable interest be collected.

Overfishing is not merely the consequence of world hunger, it is to a greater extent the corollary of the fisheries industry move from outdated cockleshell boats to factory ships, and fishing armadas of some hundred boats using sonar and helicopters, radar and electrical nets, catching too often without regard for the environment, the spawning season or the catch size. Modern trawling nets can destroy vast ocean bottom populations, thus posing a handicap for future development of fisheries.

Ocean productivity is dependent on geographical factors. In waters that are nutrient rich and favorable to photosynthetic processes, 60 kilograms per hectare may be a normal fish population. This leads to concentration of fishing activities near upwellings, in zones of turbulence where nutrients are at or near the surface near coasts and above shallow continental shelves. With the demand chiefly centered on a limited number of fish species, some small geographic areas have become the choice fishing grounds of the world. These include the Grand Banks off Newfoundland, New

England's George's Bank, the Gulf of Mexico, the Gulf of Alaska, the coast of Peru, the offshore areas of the Ivory Coast and Somali, the coast of Southwest Africa, the Atlantic Ocean north of Scotland and west of Norway, the Icelandic waters. In the latter, fishermen from England, Belgium, West Germany, Russia are common visitors. The French novelist Pierre Loti wrote a masterpiece on the Breton Fishermen of Icelandic Waters. The Grand Banks are fished by as many as fourteen nations including Japan. (Table 4)

Today it is not uncommon to sight a Soviet flotilla of twenty vessels or more, surrounding an 18,000 ton mother ship, off Nova Scotia. While the large ship speeds to the home port with at least 6,000 tons of fish, the fishing vessels keep hauling in the fish. No more than 35 million kilograms of Eastern Pacific halibut should be taken, but the limit is exceeded. The same attitude holds true for tropical tuna where a 100,000 tons limit should be respected.

Not only are "prize fish" endangered but common consumption fish like tuna in the Pacific and Atlantic, and Pacific halibut are caught in such numbers that the catch is coming dangerously close to the amount that will naturally be reproduced. English Channel herring have been drastically reduced, Moroccan coast sardines have all but disappeared, Atlantic salmon reserves have been all but exhausted by the Danish "young salmon" industry, shellfish in Micronesia are endangered by the tourist demand for large collection shells.

Yet, there are no effective regulations for sensible conservation of fish nor protection for small nations against invasion of their waters by foreign fishing vessels. A vivid example was the cod war between Iceland and Great Britain.

There is, furthermore, an incredible economic waste, an overkill in fishing. A profit would be turned out, instead of merely reaching the breaking-even point, by reducing by half the George's Bank haddock fleet, and the catch would remain identical with one fourth fewer vessels. There is an excess of capital and labor of at least $50 million in the American Northwest salmon fishing. Restrictions should not be placed so much on gear, length of fishing and size of catch, than perhaps on capital invested and labor employed.

A total overhaul of the fishing industry seems long overdue, provided it can be done equally and simultaneously in capitalist and collectivist nations.

2.2 AQUACULTURE AND AQUAFARMING

Aquaculture is the husbandry of marine species in closed basins while mariculture is the same activity in open waters. With
improved technology, both offer important new horizons, and the
Japanese are leading the way. In the U. S. since 1971 anchovies
are being successfully raised in basins at La Jolla, California.
More recently, the Fisheries Ministry of the Republic at the Ivory
Coast put special basins into use for the production of shrimp
(Penaeus ducrarum), near Abidjan, and over 500,000 shrimp are being fattened in special lagoons. In the Morbihan region of France,
similar efforts with prawns have been successful, and the production of lobsters and crayfish is expanding; aquaculture has been
improved in Brittany and Aquitaine. Commercial production of
shrimp in Japan has increased the supply by 10% which could not
meet the demand. Encouraging results have been obtained with am-
bulatory basins for tuna, but in this instance efforts will be
slowed until an international agreement is reached guaranteeing
to the producer the harvest of his "crop". (Fig. 17)

A major consideration in aquaculture and aquafarming is that
the species which are being grown find in their new environment
sufficient food and reproduction conditions which are favorable,
while not becoming competitive with the indigenous species and not
disturbing the food chain or balance on which they depend. Encouraging results were registered near Angleton, Texas, where
shrimp were raised in closed-off basins and fed a commercially
produced diet. Near Manchester in the State of Washington Coho
and Chinook salmon were raised in Puget Sound, starting from free
males' sperm. Under the Sea Grant program, high sea mariculture
experiments are conducted off the coast of Hawaii.

Since man has practically abandoned hunting in most regions
of the world as a source of nutrition in favor of animal husbandry and gathering food in favor of agriculture, would it not be
possible to consider a similar trend for the ocean environment?
Plankton yields 4,000 tons of vegetal matter for each 2.5 square
kilometers, while such a surface yields only 600 tons on land.
Naturally, since algae are the main plant life, the danger of pollution must be kept in mind and rivers and seas must be stocked,
acclimating species if necessary, in order to avoid overfertiliza-
tion which may lead to deadly eutrophication.

Danes and Zealanders have commercially caught oysters and
mussels for centuries. Today, they place the young on rafters
putting them out of reach of their carnivorous enemies and increasing fourfold their rate of growth by placing the mussel banks
in the path of faster currents. Phytoplankton production can be
intensified through a strictly supervised control, by mixing waters, and even perhaps by using the heated waters poured out by
nuclear plants. Nevertheless, the pollution factor remains

Fig. 17   Laboratory for aquaculture at Brittany Oceanographic Center.

staggering. Without it, oyster production in the United States
would easily reach 800 million tons. Some years ago 90% of the
oysters died in the Arcachon Bay, a major production center on
the Atlantic Coast of France. The disappearance of shrimp and
eels from the Belgian North Sea coast, and the decline in the
herring population, are definitely a consequence of the ecosystem's disruption.

According to Soviet scientists, the spawn of the sea urchin
has wonderful healing properties and they believe that it can
compete with ginseng, occasionally called the "root of life".
Each gram of the spawn contains as much as 35 milligrams of fat
and 20 milligrams of protein as well as an abundance of useful
microorganisms. The waters washing Kunashir and Shikotan islands, in the Kuriles, are the only place in the world it is believed, where sea urchins are available on a commercial scale.
They are caught at a depth of 80 to 110 meters by special dredges
installed on fishing vessels. The spawn is extracted from the
sea urchins, canned at the Yuzhno-Kurilsk fish processing plant,
and sent to medical institutions both within the U.S.S.R. and
abroad. Aquaculture and aquafarming are not a new activity in
the U.S.S.R. The Soviets have even recently put into service a
fish incubator. At the fish hatchery near Volgograd, white salmon are artificially bred. The male and female fish are kept in
special pools of cold water for eight months. The spawn is in-
cubated and the fry kept in covered pools with circulating cold
water and then placed in special ponds. When the fry are large
and strong enough, they are released into the Volga River and
the Caspian Sea. The complete cycle takes a year. Nevertheless,
the results are impaired by the pollution plaguing river and sea,
and by existing hydroelectric dams.

In 1976, a report publicized a two-year $415,000 contract
of California Marine Associates (Calma) and the Atlantic Rich-
field oil company. Now that it has been found possible to raise
abalone "in captivity" as it were, it may take the pressure off
the southward expanding otter population, whose appetite has been
charged with decimating the abalone industry. This contract
calls for thousands upon thousands of laboratory-bred abalone,
about 2 1/2 years old, to be planted in specially designed containers suspended beneath the ARCO Platform Holly. There the
sea delicacies will stay for from three to five years, depending
on their ultimate use, (canned, table trade, or exportation.)
If the test is successful, Calma expects to be able eventually
to utilize every oil-drilling platform in the area for the same
purpose and to market some quarter-of-a-million abalone each year.

Coho salmon, a Pacific Ocean fish, has accommodated itself
well to Maine waters. Salmon eggs bought in Washington State,

Fig. 18 Oyster beds at Arcachon, France.

Fig. 19 Marine fish farm at Andernos, France.

are hatched in shoreside fresh water tanks, transferred as finger-
lings to large nylon nets suspended in the tidal area and fed commercial salmon food enriched with ground up shrimp that colors the
fish's flesh. The salmon reaches a one-pound weight in about
half the time it would need if free.

At one salmon farm in Maine the warm water discharge of the
electricity plant provides an ideal ambient temperature by raising it 2 1/2°C in the holding pens. Hand feeding insures that
the fish get all the food they can use and eliminates waste.

Coupled with salmon husbandry is oyster production. The
first Crassostrea virginica went to market in 1976. This species
was chosen instead of the faster growing Crassostrea gigas (Japanese oyster) because of the American consumers' insistence on
"oysters on the half-shell": gigas has an irregular shell.

Six years ago, Professor Shelbourne reported raising 300,000
plaice in Ardere Loch in Scotland. He had created in the loch
conditions close to those of the ocean and showed that the death
of the fry, under natural conditions, is almost exclusively caused
by the voracity of marine flesh eaters. Marine husbandry could
thus be a remedy to this situation.

A report from the White Fish Authority of England relates
that soles were reared in tanks heated by the effluents of nuclear
power stations in Scottish lochs. The fish reached marketable
size in only one year, a process necessitating three years under
natural conditions. It is possible that fish husbandry could
even help man combat pollution. In the last five years Asiatic
Amurs have been raised in Arkansas and their incredible voracity
has cleaned up several of the state's rivers.

In some cases, however, aquaculture has not been deemed advisable. Macrocystis pyrifera   a large Pacific alga whose
stripes may reach 50 meters long; this alga grows fifteen centimeters a day and weighs dozens of kilograms. The CNEXO had planned to experiment with it at Penmarch. The alga is a rich source
of alginic acid, of medicinal value, and also used for the manufacture of oils, varnishes, electrical tape and to make tooth
imprints. However, both France and Scotland decided not to proceed because of possible damage to other underwater life. On the
other hand, the Naval Underseas Center at San Diego has proceeded
with its aquaculture at a depth of about one hundred meters.

2.3 FISH AND FOOD POISONING

Since biblical times, there have been codified injunctions
against eating certain kinds of fishes. Today, medical patholo-

gists recognize that many fishes are indeed poisonous; in fact,
some are deadly.   In 1954, Charles J. Fish and Mary C. Cobb ident-
ified 20 different families and 84 species of fishes as toxic.
Other investigators have found that of the over 1,500 species of
edible fishes in the tropics, about 300 are known to be poisonous
in certain places at different seasons.

Poisonous fishes should not be confused with fish and other
foods that become toxic through the presence and growth of dangerous bacterial organisms, such as Clostridium botulinum.   Insufficiently cooked or processed canned tuna or smoked whitefish have
resulted in cases of fatal food poisoning, due not to the inherent
poisonousness of the fish, but rather to the growth of the deadly
bacteria in it.

While in at least one case, the fish itself produces its own
toxin, a number of species acquire their poisonousness from toxins picked up elsewhere. The toxicity of these fishes can vary
a great deal and, therefore, give rise to a number of cases of
accidental fish poisoning, that is, fishes not usually poisonous
turning out to be so. Such accidental poisoning is common in the
South Pacific, with many unfortunate victims each year.   The Associated Press reported on October 6, 1947, that all of the 36-
man crew of the tugboat Edward M. Grimm became violently ill after
eating fish caught in the lagoon of Palmyra Atoll. One crewman
died and eight others were in such critical condition that they
had to be transferred to a hospital at Honolulu. On Fanning Island in the tropical Pacific in 1946-47, there were 95 cases of
fish poisoning in a population of 224 persons.

2.3.1 Suspicious Fishes

The fish poisoning above called ciguatera results from eating
the flesh of edible fishes that have been rendered poisonous by
oil-soluble toxins. Such poisoning is widespread in a tropical
belt extending around the world between latitudes 35°N and 35°S
and is encountered frequently in the South Pacific and West Indian areas. It is not confined to any single species, genus, or
family of fish, but can occur in a wide variety of unrelated fishes, including groupers, jacks, porgies, snappers, and barracudas.

The name ciguatera is believed to be of Cuban origin. It is
related to cigua, a vernacular name of the West Indian univalvu-
lar mollusc, Turbo pica, as, at first, the fish poisoning was
thought to be the same as that caused by eating the mollusc. A
Spanish dictionary of 1866 describes ciguatera as a disease suffered by a person who has eaten fish affected by jaundice or disease. It was said to cause indigestion, and persons who developed
the disease were called ciguatos or enciguatados.

The first clinical description of the ciguatera syndrome was
that of the Portuguese biologist, Don Antonio Parra, who authored
a volume entitled <u>Descripcion de Diferentes Piezas de Historia
Natural Los Mas Del Ramo Maritimo</u>, published in Havana in 1787.
It was in this account that the spelling of the name of the disease is recorded as <u>siguatera</u>. Don Antonio Parra's original description of the disease from his book is as follows:

"In the exposition of many of the fishes I have talked about,
I have said that some cannot be eaten because they are 'ciguatos'
and some others are suspicioned because they sometimes carry with
them this poison. So that the reader will become acquainted with
the nature of Siguatera, I will give a short description, although
this is probably more appropriate for physicians, since they come
in contact every day with patients which have the affliction. The
siguato fishes are those which cause symptoms similar to that of
poisoning, after you have eaten them. You feel the effects of the
poison immediately after ingestion. The initial symptoms are the
following: your color becomes pale, your features moribund, your
eyes dull, intense pains in your joints, bones, lack of appetite
decreased tactile sensation, purging, vomiting and an intense
itching all over the body; skin blemishes, ulceration, prostra-
tion and many other symptoms may be present. I can speak from
personal experience because on the 15th of March, 1786, twenty-
two of us ate a Cubera, and we all developed those symptoms to a
greater or less extent. My family became so debilitated that it
was necessary for us to obtain some assistance the following day.
All were prostrated, but each one was suffering various types of
discomfort, although the most common type of difficulty was the
extreme exhaustion accompanied by more or less pain. I took
lemonade from which I received some relief. Since we experiment-
ed with various remedies we could not determine for certain as to
what actually is the most effective treatment against the illness.
I noticed particularly that no one felt any better until they had
some sleep. In my case, I observed that I had extreme difficulty
in breathing which caused great pain and a feeling of suffocation.
My tongue became rough and I developed a sour taste in my mouth."

In a communication "On Poisonous Fishes" based on a letter
from a friend at New Providence in the Bahamas, the English philosopher John Locke (1632-1704) characterized ciguatera thus:

" ... the fish that are here, are many of them poisonous,
bringing a great pain on their joynts to those who eat them, which
continues so for a short time and at last with two or three days
itching, the pain is rubb'd off. Those of the same species, size,
shape, colour and taste, are of them poyson, the others not in the
least hurtful: and those that are, so only to some of the company.
The distemper to Men never, that we hear of, proves mortal. Dogs

and cats sometimes eat their last. In Men that have once had the
disease, upon eating of the fish, though it be those that are
wholsom, the poisonous ferment in their body is revived thereby,
and their pain is increased ..."

In the nineteenth century the principal work on ciguatera was
that of the Cuban ichthyologist Felipe Poey y Aloy (1799-1891),
entitled <u>Ciguatera Memoria Sobre la Enfermedad, Ocasionada por los
Peces Venenosos</u> ("Ciguatera, Memoir on the Disease Caused by Poisonous Fish"). In this work Poey noted that there had been enacted
an ordinance prohibiting the sale of any fish suspected of causing
ciguatera and weighing over 3 pounds. This observation is in agreement with a twentieth-century observation regarding the size of
ciguatoxic fishes.

Poey discussed the origins of poisons in nature, especially
in fishes, a subject about which he had very definite opinions.
He was the first writer to point out the rather spotty geographical distribution of ciguatoxic fishes. For example, Poey noted
that fishes on one side of an island have been observed to be
toxic, whereas, those on the other side of the same island are
often found to be completely innocuous. That this observation
was made before the significance of ecological factors was recognized is noteworthy. Poey at first considered, then rejected,
the idea that marine plants are involved in the etiology of ciguatera for many ciguatoxic fishes are known to be carnivores, exclusively. The possibility that fishes develop a condition comparable to rabies in dogs and other mammals was considered but
received little support from Poey. Other of his observations
have been proven scientifically valid, such as a recommendation
for differential diagnosis between ciguatoxic and normal fishes
by feeding the flesh of suspected specimens to mammalian experimental animals. Poey's work is considered a classic in its field.

2.3.2   Biotoxicology Studies

During the early part of the twentieth century, there was a
reexamination of conflicting data regarding marine biotoxicology.
In the early part of the fifth decade, the requirements of a war
in the Pacific caused Japanese and Western ichthyologists to concentrate independently in preparation of survival manuals for the
use of fliers forced down on remote islands and for ground troops
isolated in equally remote places. They were not outstandingly
accurate for they utilized old and poorly documented information
that had never been confirmed. More than 3,000 years after the
Biblical injunction to avoid fishes without scales and fins,
American armed forces were warned in a similar fashion.

"All important fish with poisonous flesh belong to one large

order, the Plectognathi, almost all tropical in habitat. They all
lack ordinary scales such as occur on bass, grouper, and sea trout.
Instead, poisonous fish are covered with bristles or spiny scales,
sharp thorns, or else they have a naked skin. Never eat a fish
which blows itself up like a balloon ..."

The preparation of survival manuals and recognition of the
inadequacy of available information led to a resurgence of interest in marine biotoxicology which continued and is still growing.
The outstanding scientific investigators at this time with respect to the ciguatera syndrome and the related toxin generally
are not numerous, but the number of publications issued in recent
years indicate that the problems are still troublesome.

Each year, cases of ciguatera are reported in Hawaii, Japan,
the East Indies, and West Indies. Dr. Bruce Halstead and his
group of colleagues report a mortality of from 7 to 12 percent of
those afflicted by this scourge. Complete recovery sometimes
takes many months or years.

The origin of the toxin in the different fish species still
remains somewhat obscure. One major factor about it is that its
occurrence is spotty; thus, two fishes, taken simultaneously in
the same place, may vary in their ability to transmit the disease.
One could be eaten without any ill effects and the other could
produce the agonizing symptoms of pain, spasms, nausea, cramps,
paralysis, and sometimes loss of fingernails and hair. Such extreme differences in toxicity have stimulated a widespread search
for the origin of the toxin; in the past, machineel berries,
sponges, jellyfishes, corals, sea anemones, molluscs, worms, star-
fishes, and sea cucumbers eaten by the fish have all been suspected, but careful analysis has ruled out each of these. Dr. John
Randall has speculated that blue-green algae may be the primary
source for the poison. The general consensus now is that the
toxin is something produced on the reef by some normally occurring
organism and that this organism is very small and filamentous -
a fungi, yeast, bacteria, or, perhaps, an algae. Herbivorous
fishes eat this directly and absorb the toxin; predators eat these
fishes and absorb it once again. Ciguatoxin is temperature stable,
as neither cooking nor freezing will remove or destroy the poison
in the contaminated fish.

Many authorities acknowledge ciguatera as the most widespread
disease caused by marine organisms. Mankind has been aware of
the illness for some 6,000 years, but modern medical science has
yet to completely unravel the mystery of its origin or prevent
its annual fatal occurrence.

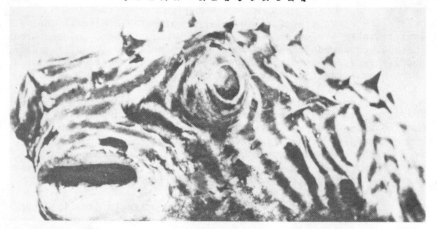

Fig. 19 A. STRIPED BURRFISH. This porcupine fish abundant
along the coasts of Florida and the Carolinas as well as
puffers and molas are tetrodotoxic. Also suspect are file
fishes, triggerfishes, spikefishes and trunkfishes.

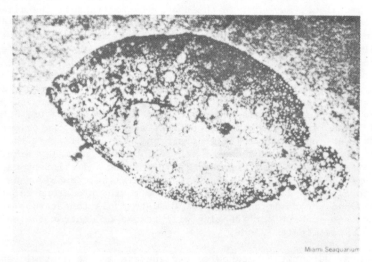

Miami Seaquarium

Fig. 19 B. FLOUNDER WITH ANISAKIASIS. Hundreds of
cases of this disease have been reported in the
last decade. Commonly called herring worm disease,
it is found in sea herring, cod, haddock and other
fish. The worms are destroyed in normal cooking,
not by light smoking, salting, or freezing.

2.3.3 Worm Infestations

In the mid-1960s, a puzzling newly discovered fish-transmitted parasitic disease began to trouble world populations. Analysis of the outbreaks of the disease in Europe and Asia traced the orign of the illness to oriental restaurants serving raw fish.

The sickness called herring worm disease or anisakiasis is a recent addition to the annals of medicine. The first recorded case was diagnosed in 1955 by Dr. E. L. Straub, a Dutch physician. Developing tastes for raw fish dishes in oriental restaurants has helped to increase the numbers of victims of this disease which, by 1965, was recognized as a world problem.

Sashimi, gravlox, and smoked fish serve to spread the illness. The parasite which is found in saltwater fishes, is killed by heat and destroyed by freezing. It will survive in vinegar for about 50 days and will occur in smoked fish because insufficient heat is produced to kill the worm.

In Holland, infection developed in persons who ate Dutch Green Herring, a variety of raw lightly salted sea herring. The sickness is a result of small inch-long worms causing an inflammation of the bowel wall. The Dutch government stopped the disease in Amsterdam by passing legislation requiring that commercial herring be frozen for at least 24 hours before marketing. Fishes such as flounders, cods, and haddocks have all been recorded as hosts to the particular parasitic worm causing anisakiosis. Cases of the disease have been reported throughout the world, including the United States. In March, 1974, the first North American symposium on anisakiosis was held at Rockefeller University at New York City. At that time, the extent of herring worm disease as a world problem was still being discussed, but it was concluded that elimination of the disease could be achieved by prohibiting the eating of uncooked fish.

Another world-wide fish-transmitted disease is the broad fish tapeworm, Dibothriocephalus latus. This tapeworm, which parasitizes humans, comes from eating insufficiently salted or cooked pikes, carps, whitefishes and other freshwater species that harbor the cestode worm. Human infections by this fish worm have been recorded by physicians in Finland, Ireland, Israel, central Africa, Siberia, Japan, Chile, Canada, and in the central United States surrounding the Great Lakes. In Finland, about 20 percent of the population is infected by this parasite.

Symptoms of this disease, called dibothriocephaliosis, include abdominal pain, loss of weight, and pernicious anemia. Anemia develops because this species of tapeworm has a special affinity in its human host for vitamin B-12 and results in retarding blood cell formation.

One of the most widely endemic diseases in Asia, affecting
an estimated 19 million people in China, Japan, Korea, Vietnam,
and India, is spread by carps and other cyprinid fishes. The
disease, called opisthorchiosis, is caused by the parasitic trema-
tode liver fluke, Opisthorchis sinensis. When the infected fishes are eaten by human beings, the worms spread to the liver and
bile ducts where cirrhosis and ultimately death can result. Well-
cooked fish does not transmit this severe illness.

There are, as well, several other fish toxins and fish-transmitted diseases, most restricted to a few particular groups. Some
herrings contain a toxin, some tunas contain another, and so on.
The only way to effectively guard against such diseases is, when
eating locally caught fishes (canned and processed fishes generally can be considered safe), to check with the locals about the
possibilities of poisoning, by being at least somewhat familiar
with the fishes likely to be involved and, especially in cases of
parasite infestation, by cooking thoroughly or freezing any fish
to be eaten.

2.4 HUNTING

Environment and conservation groups have been successful in
putting some countries out of the whaling business and prohibiting
the import of products of cetacean origin by the United States.
The making and use of fur coats is now under attack by environmentalists. Seal and sea lion hunting are in great popular dis-
favor. Some voices even object to the training of porpoises for
military aims. Efforts are also being made to avoid the killing
of dolphins caught in the tuna fish nets. In this connection it
is worthy to note that a large international outcry was raised
against Japanese fishermen who killed large numbers of dolphins
in 1977 and 1978.

Because tuna schools often swim underneath large groups of
dolphins, fishermen devised nets to catch them together. Dolphins jump over the net, but many dive instead, and being mammals,
consequently drown for lack of air when caught in the nets. An
estimated 100,000 dolphins were thus killed in 1975, which led
to the outlawing of such nets in 1976.

The hunting of sea mammals appears to be declining since
only Australia, Japan, and the U.S.S.R. continue large scale
operations. On the other hand, some of these mammals are being
trained to help man retrieve objects lost at sea while others aid
in the study of sound transmission in the oceanic milieu. Norway
which had interrupted whaling resumed limited activity in 1975.

Although emphasis is usually placed on whales, the list of

sea mammals is quite long. Many of them were hunted for centuries
for their fur, for oil, and as a food source for Arctic populations. The Blue Whale (Balaenoptera musculus) largest mammal ever
to live on earth, now an endangered species, inhabits all oceans.
Other mammals, however, are regional. They include seals, porpoises, dolphins, sea otters, manatees, walrus, narwhals, and, to
some extent polar bears. At least 46 different mammals inhabit
the waters of the Western Hemisphere. (Fig. 20)

In Tomorrow's Wilderness, published several years ago by the
Sierra Club in San Francisco, Athelstan Spilhaus recalls the story
of the fur seals of Pribilof Islands in the Bering Sea, "Here 80%
of all fur seals of the world are born. When the islands were
discovered, 200 years ago, there were 2 1/2 million seals there,
but by the turn of our century killing on land and wasteful seal-
ing at sea had eliminated sixteen out of every seventeen seals!
The herd was almost extinct! Since then, with a series of international agreements to protect it, the herd has grown back to
its present strength of 1 1/2 million!"

The International Whaling Commission helps in reducing the
slaughter by setting limits on the length of the hunting season,
the catch, the minimum size and forbidding the capture of nursing
mothers, calves, right whales, and Pacific gray whales.

Unfortunately quotas have been repeatedly set too high, re-
gardless of the advice of the I.W.C. experts themselves. In the
1963-1973 decade, only once did whalers manage to fill the quota
set! Biologists estimate that at best 2,000 blue whales (Balaenoptera musculus) and 50,000 whales (Balaenoptera physalus) are
left. There were close to 29,000 blue whales caught in 1931, by
1965 the catch was 66; between 1965 and 1962, the average yearly
catch of fin whales reached 25,000; since 1965 this average has
dropped to 2,000, and is steadily declining since 1973. Sei whales
were hardly hunted prior to 1946; the kill of Balaenoptera boreal-
is peaked in 1966 with 19,000 animals; by 1969 only 5,000 were
caught, and the number is rapidly declining.

Individual species quotas have only been set by the I.W.C.
since 1972, when an end was put to the system of Blue Whale Units,
which permitted capture of any species on the basis that 1 blue
whale unit = 1 blue whale = 2 fin whales = 6 sei whales. This
was a comparative oil yield measure.

3. RAW MATERIALS

Marine resources, we pointed out, are either chemical, geological or biological. Geological resources can be either authigen-
ic, detrital, or organic. Such resources occur in large quantities

WHALES

RIGHT WHALE

The right whale is an almost-extinct member of the baleen family, whales with plates instead of teeth in their mouths Up to 60 feet long, it was named by whalers who thought it the right whale to kill because it swam very slowly and floated when dead Fewer than 500 remain (Balaena glacialis)

PILOT WHALE

Thousands of these toothed whales—which grow to a length of 22 feet—are killed each year for their meat and oil. Pilot whales travel in a large group, all of which will sometimes beach themselves on the sand and die. The suicidal action remains a puzzle to scientists. (Globicephala melaena)

KILLER WHALE

One of the fastest and most intelligent whales in the ocean, the toothed killer whale grows to a length of 30 feet, half that of the sperm whale. Its main diet consists of fish, but it was named "killer" because it also eats meat, including other whales, dolphins, seals and sea lions. (Orcinus orca)

FIN WHALE

This baleen whale is the second largest, up to 80 feet in length vs. the blue whale's 100 feet. Fin whales hunt food by herding small schools of tiny fish into a ball and scooping up the ball in their enormous mouths. So many fin whales are being killed that scientists fear for their survival (Balaenoptera physalus)

HUMPBACK WHALE

Most playful of the big whales, the humpback loves to leap into the air and bellyflop with a crashing splash. This 50-foot-long member of the baleen family also loves to sing, and, in one place, Bermuda, it sings different songs every year. No one seems to know why these whales do it. (Megaptera novaeangliae)

BOTTLENOSED DOLPHIN

This dolphin (growing up to 8 feet in length) is the one you see most often doing tricks in oceanariums. When the dolphin does something he likes, the trainer (the trainee, from the point of view of the very intelligent dolphin) blows a whistle and rewards the performer with a fish (Mesoplodon bidens)

FIG. 20

WORLD FISHERIES

WORLD COMMERCIAL CATCH OF FISH, CRUSTACEANS, MOLLUSKS, AND OTHER AQUATIC
PLANTS AND ANIMALS (EXCEPT WHALES AND SEALS), BY COUNTRIES, 1972-76

Country	1972	1973	1974	1975 (1)	1976
	- - - - - - - - - Thousand metric tons - - - - - - - - -				
			Live weight		
Japan.	10,272	10,748	10,805	10,524	10,620
U.S.S.R.	7,757	8,619	9,236	9,936	10,134
Peoples Republic of China					
(Peking).	(2)6,880	(2)6,880	(2)6,880	(2)6,880	(2)6,880
Peru	4,725	2,328	4,145	3,447	4,343
Norway	3,186	2,987	2,645	2,550	3,435
United States.	(3)2,695	(3)2,719	(3)2,744	(3)2,743	(3)3,004
Republic of Korea.	1,341	1,684	2,023	2,133	2,407
India.	1,637	1,958	2,255	2,328	2,400
Denmark.	1,443	1,465	1,835	1,767	1,912
Thailand	1,679	1,679	1,516	1,553	1,640
Spain.	1,536	1,578	1,510	1,523	(2)1,483
Indonesia.	1,270	1,265	1,336	1,382	1,448
Philippines.	1,128	1,251	1,298	1,366	1,430
Chile.	818	691	1,158	929	1,264
Canada	1,169	1,157	1,037	1,029	1,136
South Vietnam.	(2)978	(2)1,014	(2)1,014	(2)1,014	(2)1,014
Iceland.	726	902	945	995	986
Brazil	602	704	765	(2)836	(2)950
France	797	823	808	806	806
North Korea.	(2)800	(2)800	(2)800	(2)800	(2)800
Poland	544	580	679	801	750
Bangladesh	(2)640	(2)640	(2)640	(2)640	(2)640
Republic of South Africa .	664	710	648	636	638
Namibia (S.W., Africa) . .	(2)527	(2)710	(2)840	(2)761	(2)574
Mexico	459	479	442	499	572
England and Wales.	539	557	534	497	520
Malaysia	359	445	526	474	517
Scotland	530	562	538	468	514
Burma.	453	463	434	485	502
Nigeria.	446	466	473	478	495
Federal Rep. of Germany. .	419	478	526	442	454
Italy.	430	399	431	417	420
Senegal.	294	316	357	363	361
Faeroe Islands	208	246	247	286	342
Portugal	452	482	436	375	339
Netherlands.	348	344	326	351	284
Argentina.	238	302	296	229	282
All others	7,132	7,377	7,172	7,150	7,171
Total	66,121	66,808	70,300	69,893	73,467

(1) Revised.
(2) Data estimated by FAO.
(3) Includes the weight of clams, oysters, scallops, and other mollusk shells. This weight
is not included in other U.S. catch statistics.

Source:--Food and Agriculture Organization of the United Nations (FAO), Yearbook of Fishery
Statistics, 1976, Vol. 42.

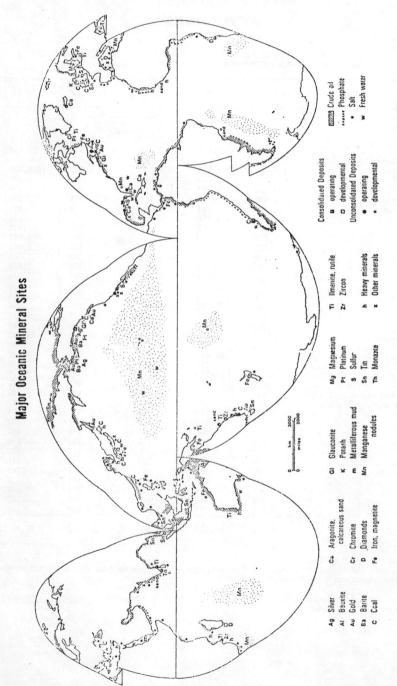

Major Oceanic Mineral Sites

Compiled by: R.H. Charlier

Consolidated Deposits
◼ operating
◻ developmental

Unconsolidated Deposits
● operating
∙ developmental

▨ Crude oil
··· Phosphate
• Salt
w Fresh water

Ag Silver
Al Bauxite
Au Gold
Ba Barite
C Coal

Ca Aragonite, calcareous sand
Cr Chromite
D Diamonds
Fe Iron, magnetite

Gl Glauconite
K Potash
m Metalliferous mud
Mn Manganese nodules

Mg Magnesium
Pt Platinum
S Sulfur
Sn Tin
Th Monazite

Ti Ilmenite, rutile
Zr Zircon
h Heavy minerals
x Other minerals

km 3000
miles 2000

but specialists find that some are and some are not profitably
exploitable.

3.1 THE ECONOMICS OF OCEAN MINING

In 1963, the February issue of <u>International Management</u> in a
paper entitled "Prospecting the Deep" announced that an American
company was drilling in the Gulf of Mexico at 100 meters for petroleum, that the Japanese had begun large scale exploitation of
the sand-iron deposits on the bottom of Ariake Bay and that the
depots of the Indian Ocean were being raked for "minerals-contain-
ing lumps". It was then estimated that metals could be obtained
from the sea at only 50 to 70% of the cost of land mining, a pro-
position made the more attractive because on land high grade ore
is being rapidly depleted and marine ores are often highly concentrated. Realistic estimates list no less than 60 useful elements in the oceans, but even though the quantity present is often
tremendous, such as 10 million tons of gold, over 15 billion tons
of manganese and at least 20 billion tons of uranium, these elements occur in billions of tons of water. The ocean floor offers
greater concentrations, and the continental shelf is a repository
of substances such as bromide, magnesium, tin, iron, phosphorite,
sulfur, and in localized areas diamonds, gas, and petroleum. The
abyssal zones are covered with a red clay containing approximately 50% silica, 20% aluminum oxide, in addition to manganese.
cobalt, copper, nickel, and vanadium.

In 1973, the world was stunned by the oil embargo and sud-
denly marine minerals became far more important to the economies
of the world. The United States was faced with an increasing de-
pendence on foreign supply sources for 31 minerals, including
sand, gravel, copper, nickel, and uranium. While imports cost the
United States $6 billion in 1971, they are expected to reach, in
steady dollars, $50 billion by the year 2000.

Ocean mining is complex, difficult and costly but the 1975
report of the National Research Council (NRC) nevertheless con-
cludes that "marine mining offers enormous potential for becoming
independent of foreign countries for some important minerals, including those used as a source of energy". Evidently, this reality has not escaped other nations either, and the desire to carve
up the ocean into national bailiwicks is certainly motivated by
such considerations.

Careful attention has been paid to the potential environmental impact, and, according to the 1975 report, marine mining
could be engaged with environmental risks, although the NRC insists on the need to set standards before exploitation. It pro-
poses a regulatory system which would prohibit industry from

gaining economic advantages by keeping for its private use information it has amassed in prospecting. Instead of area-leasing used
for offshore oil extraction, work program proposals would be the
basis for allocations. As mining activities must be integrated
into coastal zone planning, licensing schedules should be on a
ten-year schedule to insure orderly procedures.

On the continental shelf the probability of mining operations
are classified as follows in order of earliest need: sand and
gravel, calcium carbonate, titanium and gold placers, phosphorite.
In the United States this would mean mining in the Gulf of Maine,
along the Massachusetts coast, the New York - New Jersey bight,
the Northwest and Southwest Pacific coasts, the Bering Sea, the
Arctic Shelf, and, inland, along the Great Lakes. Based on land
experience, a span of at least seven years must be anticipated
between exploration and full production.

As will be discussed later, deep ocean mining is entangled
in a web of international legal maneuvering which already has en-
gulfed the three United Nations "Law of the Sea" conferences of
Caracas, Geneva and New York (1974-75-76) and two Pacem in Mari-
bus convocations (Malta 1974, Japan 1975). Nevertheless, it is
expected that the United States will unilaterally legislate guar-
antees within the next two years enabling deep sea mining to start.
Much is expected from the mining of deep sea polymetallic nodules
which contain high grade manganese, copper, cobalt and nickel.

Yet, deep ocean mining on a production scale requires a
staggering initial capital investment. Speaking in 1975 dollars,
between $240 and $900 million would be needed to extract manganese,
copper and cobalt near Hawaii and process 5000 tons a day in California, while requiring perhaps 20 years for full production.
This seems to point to the need of governmental incentives to encourage capital investment in ocean mining. But besides money
further refinement of technology must be encouraged and manpower
appropriately trained, the more so since no university at all offers a formal degree in marine mining.

Basic research is steadily conducted, but the applied aspects are often embodied in projects. For example, Project FAMOUS
(French-American Mid-Ocean Undersea Study) aimed at getting further understanding of sea floor spreading and at probing the mid-
Atlantic rift valley where earth crust is created, has examined
the creation of such minerals as copper, manganese and chromite.
One theory holds that sea water circulating through fractures in
the ridge's rock formations may carry off some of these materials
and concentrate them elsewhere. Perhaps the study may give a lead
to minerals siting in more easily accessible locations, thus increasing prospecting's efficiency and rationality. The submerged
extension of the continent off the Atlantic, Pacific and Gulf

coasts of the United States has been claimed by the United States
for economic exploitation, the most extensive territory to be
added to the country since the Louisiana Purchase in 1803. The
Continental Shelf extends around the United States from 10 to 300
miles off the coast including 175 miles off Cape Cod, from 50 to
125 miles off the South Atlantic States, from 50 to 150 miles
into the Gulf of Mexico, from 10 to 50 miles off the Pacific Coast,
and approximately 300 miles off the Alaskan coast. The Hawaiian
Islands' shelf extends 10 to 50 miles offshore.

Already underway in various parts of the world, efforts to
extract wealth from beneath the sea include: extensive recovery
of oil off the shores of the U. S.; diamond mining off the coast
of southwest Africa; iron and coal mining off the Continental
Shelf of Japan and tin off the Malaysian Shelf; and the extraction of magnesium and bromine from the sea at Freeport, Texas.
(Fig. 21)

3.2 THE PRODUCTS

Ten years ago, as shown in Table 5, the value of mineral offshore production along U. S. coasts alone reached the equivalent
of roughly 7.3 billion current dollars.

3.2.1 Phosphorite

Clumps of phosphorite appear often near shore at less than
100 meters depth. Phosphorite leases were granted for its mining
over 12,000 hectares near Southern California. In 1963 it was
estimated that the fertilizer could be placed on the market at
$13.50 a ton, while $15 was the price for the imported product.
In 1973, the phosphorite mined on land in coastal North Africa and
Florida was worth $13, and a ton of the marine product cost only
$6. With the Arab producer heading a pricing organization, one
ton is priced at more than $17 since 1976 and may still increase.
Phosphorite nodules are also available at great depths, but they
are as difficult to retrieve as manganese nodules.

Phosphorite contains usually 30% of economically worthwhile,
material (P_2O_5) and is generally found where other materials
brought from land have not accumulated. Exploitable areas include
the Pacific coasts of both California and northwest Mexico, the
Atlantic coasts of Georgia and Florida, Peru, the west coast of
South America, Japan, Spain, the northwest and the south of Africa,
and probably the northwest coast of Australia as well. The Indian
geological survey claims to have found deposits near the Andaman
Islands. The composition of phosphorite is sand, gravel, calcareous organic remains, and fossil phosphorite.

Fig. 21 RECENT MARINE MINING ACTIVITIES

Location	Activity [1]	Depth (feet) [2]	Interest
AFRICA			
Red Sea	Exploration	6,000±	Sulfide muds.
Southwest Africa	Dredging	600–	Diamonds.
Union of South Africa	Exploration	(?)	Phosphate.
ASIA			
Borneo	Exploration	600–	Tin.
India	Exploration	(?)	Phosphate.
Indonesia	Dredging	150–	Tin.
Japan	Dredging	30–	Iron sands.
Malaysia	Exploration	600–	Tin.
Papua and New Guinea	Exploration	600–	Iron sands.
Philippines	Exploration	600–	Iron; gold; titanium.
Thailand	Dredging	150–	Tin.
EUROPE			
Iceland	Dredging	150±	Shell sands.
Great Britain	Exploration	600–	Tin.
NORTH AMERICA			
Bahamas	Dredging	150?	Aragonite.
Canada (B.C.)	Exploration	12,000±	Manganese nodules.
Caribbean Sea	Exploration	200–	(?).
Mexico	Exploration	600–	Phosphate sands.
Pacific Ocean	Exploration	12,000+	Manganese nodules.
U.S.A.:			
Alaska	Exploration	200–	Gold.
Blake Plateau	Exploration	600–2,400	Manganese; phosphate.
California	Dredging	30±	Shells.
California	Exploration (inactive)	600±	Phosphate.
Louisiana	Mining (Frasch)	60±	Sulfur.
New England	Dredging	300–	Sand.
North Carolina	Exploration	600–	Phosphate sands.
OCEANIA			
Australia	Exploration	600–	Phosphate; heavy metals.
New Zealand	Exploration	600–	Heavy metals.
Solomon Islands	Exploration	600–	Tin.
Tasmania	Exploration	600–	Heavy metals.

[1] Dredging operations generally include exploration activity. Does not include mines originating on land and drifted out under the sea floor.
[2] Less than is represented by –; more than is represented by +; approximately is represented by ±.

Source: Department of the Interior.

Table 5. Value of offshore (U.S. coasts) mineral production
(1960-66) in millions 1968-dollars (roughly 20%
below 1976-dollars)

Years	Magnesium metal and compounds, salt & bromine (sea water)	Well-petroleum, natural gas and sulphur (ocean subfloor)	Sand, gravel, zircon, feldspar, cement rock, limestone (beaches & seafloors)	Totals
1960	69	423.6	46.8	539.4
1961	73	496.6	46.2	615.8
1962	89.1	620.7	44.3	754.1
1963	84.6	730.8	42.5	857.9
1964	94.5	820.3	43.6	958.4
1965	102.6	933.3	51.4	1,087.3
1966	117	1,777.7	49.2	1,343.9
Totals	629.8	5,203.0	324.0	6,156.8*

* 7300 (approx. 1976-$)

Source: U. S. Dept. Interior, Bureau of Mines

World reserves of phosphorus considerably exceed the projected cumulative demand. Even rapid expansion of agricultural demand would not greatly reduce the reserves. Land production, however, contributes to pollution, is in conflict with conservation measures and, because of bulk, results in high transportation costs. These later considerations are the major factor for looking towards the sea. Cumulative world demand for phosphorus is estimated at 200 million tons, but reserves reach 21,500 million tons. (Fig. 22)

3.2.2 Polymetallic nodules (Mn, Cu, Co, Ni)

Nodules containing a high concentration of manganese have been found in many ocean areas of the Pacific, the Atlantic, and the Indian oceans. Pacific nodules contain 30% of manganese, plus copper, colbalt and nickel. Usually less than 20 centimeters in diameter, one nodule weighing 750 kilograms and measuring one meter in diameter was brought up by the British from the Philip-

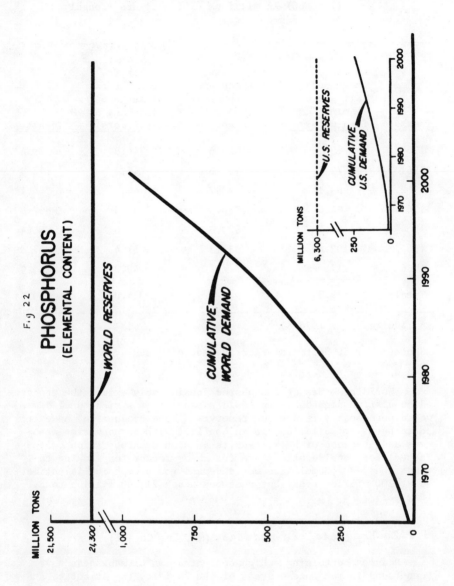

Fig 22

PHOSPHORUS
(ELEMENTAL CONTENT)

Fig. 22B PROJECTED DEMAND FOR GIVEN MINERALS
TO 1985 AND 2000

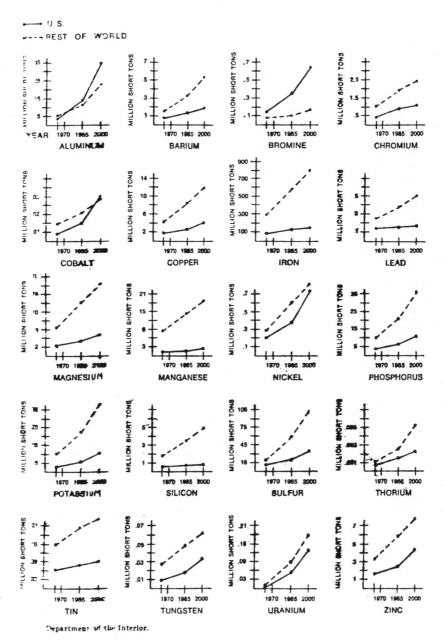

Department of the Interior.

pine Trench. In 1970, an area covering 390 square kilometers, in
waters of medium depth, was located near Hawaii. Heaviest concentrations were found near the north coast of Lihue and south
Kapaa. A French expedition picked nodules off Tuamotu (Tahiti)
from depths varying between 1,000 and 1,600 meters. Most samples
were small but one specimen weighed 128 kilograms. A team of
oceanographers from Columbia University recently made a map of
the manganese nodule sites which might well prove to be a valuable
part of the economic atlas of the ocean referred to earlier.

There are several theories concerning the origin of the polymetallic nodules. One theory is that they are from precipitation
of sea water particles originating from terrestrial and sub sea-
bed sources, and undersea volcanic eruptions. Another theory
views the nodules as organic phenomena with living organisms as
agents through which metals are deposited on a nucleus. Still
another theory, while admitting that nodules could occur without
bacterial intervention, holds that bacteria produce a catalyzing
enzyme helping the reactions occurring at deep ocean pressure and
temperatures. (Fig. 23)

Economically most interesting, because of high ore grade
and favorable weather conditions, are the nodules southeast of
Hawaii along the Pacific equatorial belt, around 9 degrees of
latitude.

It is already more than a hundred years since the Challenger
reported the presence on the ocean bottom of nodules containing
manganese. In 1968, an easily exploitable deposit was located
under the waters of Lake Michigan, spreading over 500 square kilometers and at depths varying from 30 to 60 meters. The yield was
assessed at 40,000 to 60,000 tons per square kilometer. These
nodules are actually made up of several metals and non-metals. A
partial list, in order of decreasing importance, includes manganese, iron, aluminum, nickel, copper, and cobalt. The American
concern Teneco in 1969 announced plans to gather these nodules.
The operation was actually to be carried out by its subsidiary,
Deep Sea Ventures, which had gathered 40 tons of nodules off the
Florida and Carolina coasts. Ocean Resources had also been active
in this field. So have Canadian, German, Japanese and French
corporations. Howard Hughes Tool Co. built a barge as large as
a U. S. football field which is submersible. It left San Francisco in January, 1974, for South America in search of undersea
mineral resources, and particularly polymetallic nodules. Bethlehem Steel has also been involved for some time in exploitation
studies. Summa Inc., a Howard Hughes subsidiary, is engaged in
plans for manganese mining; so is Kennicott and foreign firms
such as Le Nickel, Sumitomo, metallgesellschaft A.G., Preussag A.
G. and others. International Nickel has supported research in
the field, on processing technology and studies in engineering

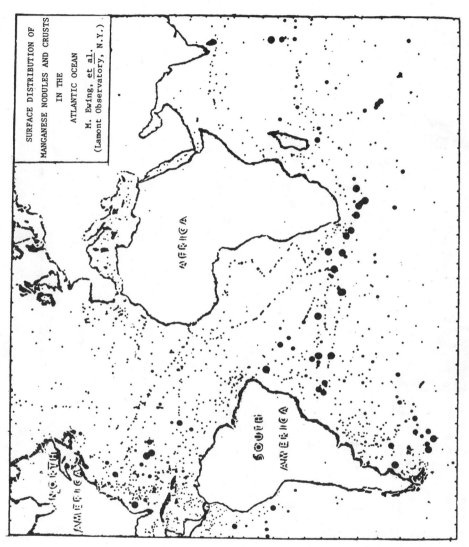

Fig. 23a
SURFACE DISTRIBUTION OF MANGANESE NODULES AND CRUSTS IN THE ATLANTIC OCEAN

legend   fig.23 a and b   Manganese ✦   · 0% - 25%
                                                    • 25% - 50%
                                                    ● 50% - 75%
                                                    ● 75% -100%
      ✦ Proportion of ocean floor covered by survey. Covered stations
        occupied from August 1963 to January 1970

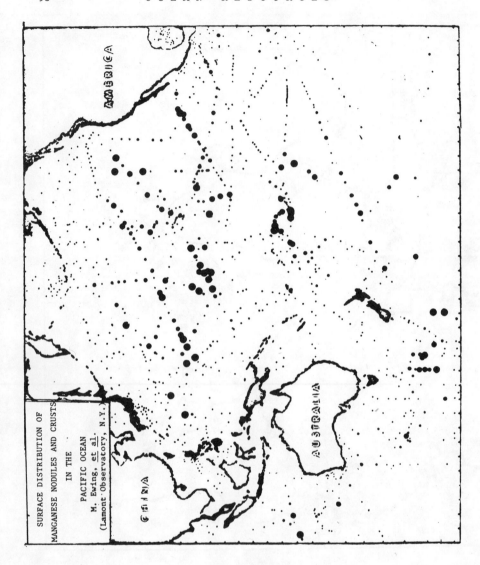

Fig. 23b

SURFACE DISTRIBUTION OF MANGANESE NODULES AND CRUSTS IN THE PACIFIC OCEAN

systems.

Ferrous-manganese concretions have been sited in the Sea of
Japan and chromite could be exploited along the Sakhalin coast.
According to John Mero, the Pacific Ocean alone has reserves of
10^{12} tons of manganese and a single gathering operation could provide up to 50% of the world's production of cobalt. P. L. Bez-
rukov of the U.S.S.R. Academy of Sciences Oceanology Institute
describes the 1969 to 1970 <u>Vitiaz</u> campaign as follows: "Nodules
of manganese are found at depths of 4 to 6 kilometers at rather
large distance from continental masses in regions of accidented
relief and slow sedimentation. The sediments are red clays and
diatomaceous or radiolarian oozes. At lesser depths, 1 to 2
kilometers, nodules rest upon igneous or carbonaceous rocks and
upon the ocean bottom. The highest density is 50 to 75kg/m2 in
the central Pacific where phosphatic rocks are abundant on the
slopes of submarine mountain chains which stretch westwards from
Hawaii." These observations conform to the photographic reconnaissance made by American scientists. If the Soviets are right,
then reserves of over 100 billion tons of cobalt, manganese, and
nickel rest on the bottom of the Pacific Ocean. If all this data
is correct, then less than 1% of ocean bottom reserves would
suffice to satisfy current needs in manganese, nickel, copper
and cobalt for 50 years. According to Brooks, the price of manganese production could drop by 45%, that of nickel by 7%, and that
of cobalt by 30%. His views are challenged by Sorensen and Mead
who see only reductions of 3% and 4% for manganese and nickel,
respectively, but still 27% for cobalt. (Fig. 24, 25)

According to D. S. Cronan of the University of Ottawa, manganese at depths exceeding 3,300 meters is mostly todokorite, while
at lesser depths it is manganese dioxide. Todokorite concentrates
nickel and copper which replace the bivalent dioxide which concentrates, instead, cobalt and lead. Todokorite is more common in
an oxidation milieu.

Actually world reserves of manganese can carry us through
this century. As an ingredient of steel it may become more valuable if developing nations accelerate their industrialization
pace. But there is precious little manganese in the United States
which imports virtually all it needs. Considering that as much
as 40% of the price paid is for transportation, the interest of
the United States in the nodules is well placed. (Fig. 24)

According to figures released some years ago, the cumulative
world demand is expected to reach 200 million tons in 1980, with
world reserves assessed at 480 million tons. The current U. S.
reserves may total 6 million tons, but the demand is at least
three times as much.

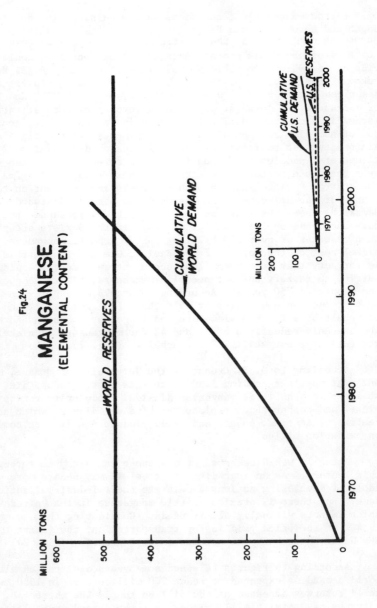

Fig.24

MANGANESE
(ELEMENTAL CONTENT)

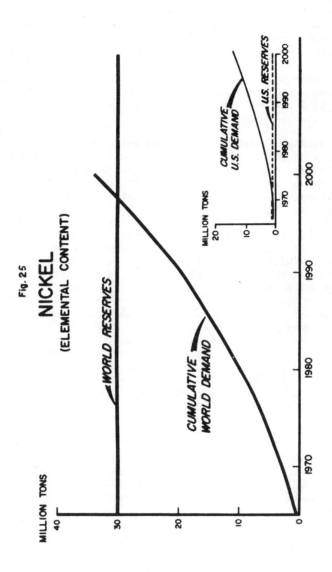

Fig. 25

NICKEL
(ELEMENTAL CONTENT)

The total value of manganese nodule mining activity including onshore and offshore activities has been estimated at $600 million. From an economic viewpoint, ocean manganese and cobalt can be considered worth at least the value of their land counterparts which they would replace, and even more if the lower prices were to permit increased consumption.

If only nickel, cobalt and copper were extracted, the mining operation would still yield a small profit, although cobalt production would become unprofitable if costs of processing nodules would exceed $45 per ton of nodules.

Only a small percentage of sea bottom has been surveyed but knowledge is sufficient to start appreciable exploitation and the U. S. has the most advanced technology. Most attention, as shown above, has been centered on North Pacific nodules found at 3000 to 4500 meters. Four companies have announced they are ready to begin mining.

If the proposed Exclusive Economic Zone (E.E.Z.) is created, polymetallic nodules will virtually be the only mineral resources. exploitable within the relatively near future in what will be left as international waters. (See V). No significant recovery of nodules can be expected within the E.E.Z. itself, the more so since economics will force oil and gas exploitation to the shallower waters even though technology would allow deeper water recovery. Many countries do not at this time need some of the metals contained in the nodules. Indonesia, the Philippines, Australia, Canada, the USA, the Dominican Republic, and Guatemala, have ample supplies of nickel on land. The cost of transporting nodules to processing sites is three to four times as much as the price for land ores, though cost differentials could be compensated by copper recovery from the nodules, while copper is not present in lateritic ores. Processing at sea appears impractical at the present time.

In terms of dollars, we must thus consider that 66% will be added to the cost of the product to recover the nodules, and bring them to a plant. This can be estimated as representing an outlay of capital of $5 per pound per year, or roughly $150 million for $30 million of nickel.

3.2.3 *Iron Ore*

Current value of dredged and mined subsea minerals exceeds $600 million annually, but accounts for only 2% of world production.

In 1961, the Japanese started retrieving small quantities of
iron ore from the bottom of Ariake Bay. With the help of two giant
dredging units production rose by 1963 from 1,000 to 30,000 tons
a month. Yawata Steel estimated at that time that it had a minimum of 36 1/2 million tons available in reasonably shallow waters.

Construction is under way in the Maritime Territory district
of the U.S.S.R. of a metallurgical complex which will process manganese, cobalt, nickel, and copper-containing materials of marine
origin. The plants will treat magnetite - and titomagnetite-rich
placers mined from the Baltic and Black seas. Other such deposits
have been located in the Sea of Azov and near the Kurile Islands.

3.2.4 Tin

Among Soviet plans are recovery of tin from the Laptev Sea
and the east Siberian costs, and of amethyst along the southern
littoral of the White Sea. Cassiterite, a tin ore, has been mined from the ocean for some time: in the State of Selangor, Malaysia, tin is mined from 50 meter depths. The 65th anniversary of
tin mining near Tongkah, Thailand was celebrated in 1972. Deposits of cassiterite were found in the Andaman Sea near Takuapa, and
exploitation started there in 1968. In Indonesia, eluvions and
alluvions containing tin ore are dredged from the ocean bottom,
close to the isles of Singkep, Banka, and Billiton. In 1971, an
American concern tested a hydraulic mining system which proved
functional at depths down to 1,000 meters. The largest tin ore
dredge was built recently by the Japanese. They scrape the ocean
bottom then suck up the material made up of sands and muds. Once
aboard the ship, the ore is automatically separated from the bulk
of the material which is then returned to the ocean. Japanese
experiments are underway with a continuous belt dredge which could
perhaps work at depths of up to 4,000 meters.

The Soviets reported large tin reserves in the Vankina Guba
(Yakutia), near Selyakhskaya, stretching from Cape Svyatoi Nos, to
the Strait of Dmitri Laptev, south of Bolshoi Lyakhov Island, and
in the harbors of Khuntzeyev and Siahu, and also in the Japan Sea.
They claim to be ready to extract diamonds, platinum, and especially gold near Kalyma (Lena River), Nakhodka (Okhotsk Sea) and along
the northern and southern coasts of the Kamchatka Peninsula.
Specially equipped ore processing ships are being built. They
will also serve for prospecting and be equipped with a probe with
a radioisotope sensor which, when put to sea, will reveal on a
shipboard indicator the presence of lead, gold, and manganese
through gamma ray absorption.

In the Atlantic Ocean, off the coast of Cornwall, in St. Ives
Bay, recovery of tin-containing sands has started and important

deposits have been located off Brest, on the coast of Brittany.

3.2.5 Sands and Gravel

Though some scientists doubt the development of a large ocean
mining industry, ocean mining already is a reality. Sands and
gravels are being extracted from the ocean. Some, containing organic remains, have been used as building stone; San Marcos Castle,
in St. Augustine is built of coquina. Sands and gravels are used
for artificial beach building, land fill, cement production and
in the making of prestressed concrete. These materials are in-
expensive and easily transported.

Ocean Cay is an artificial island built in the Bahamas Archipelago using dredged material. Aragonite is sucked up and treated
on this island; production reached two million tons during 1971
and reserves are estimated at 575 million tons. Near Muiden, in
The Netherlands, sand is dredged from a depth of 75 meters under
the surface of the former Zuiderzee. Off the British Isles, more
than 50 dredges exploit sand and gravel deposits. Recently, high
quality calcareous sands have been located near the Laccadive
Islands. These mining operations may prove dangerous if carried
out too close to the coasts as the Lebanese and the Israelis
found out when beaches were ruined. Yet, the need for marine sands
and gravels will increase. Some countries fail to find sufficient
quantities on land, and the French foresee that within a decade
the Channel will be tapped for the materials needed by Paris, Nor-
mandy, and the north of France.

It is generally more expensive to recover materials offshore
than on land; however, transportation costs may be less, hence the
competitiveness when the user is located on or near the coast. Inland locations may be served at low cost if materials can be transported by water. Furthermore, depletion of land-based supply centers may act as a favorable factor for offshore site development.

Depth is an important element in gravel development. Econi-
cally exploitable deposits should not lie beneath more than 35
meters of water. Emery believes that U. S. Atlantic and Gulf of
Mexico continental shelf deposits would satisfy U. S. needs for
centuries to come. Some 450,000 millions of tons of sand are
available along the Northeastern coast of the United States but
these reserves have been tapped so far only for small landfill
and beach repair operations.

Gravel is more difficult to obtain because it is often buried
under thick sand deposits. In the United States rich deposits
line George's Bank and the areas off New York City. Calcareous
marine shells, abundant along the Gulf of Mexico have been mined

along the coasts of Louisiana, Texas, Florida and along the Atlantic and Pacific shores yielding 20 million tons a year. Objections to sand and gravel dredging have been voiced by environmentalists. Rich U. S. deposits exist in Hawaii, and the Caribbean also has huge reserves.

Some sands contain gold. Gravels contain diamonds and often where gold is found there is also platinum. Shell Oil is prospecting off Alaska. Goodnews Bay has provided since 1935 almost 90% of the platinum needed by the United States. Extraction is presently carried out on Australian and South African beaches. Here diamonds have been mined from the sea since 1962 along the coast from the Orange River mouth to Deay Point. Kimberlite is transported by the river and deposited along the littoral by marine action. The yield is not negligible; in 1964, one company extracted 16,118 carats from a single marine deposit. In Alaska, Inlet Oil exploits petroleum deposits, but also discovered gold deposits in the channel off Bluff and continues its search in Goodnews Bay for gold, platinum, mercury, and chrome, and in the southeast for gold, silver, copper, zinc and uranium. Beyond the Burdekin River in Queensland, Australians have found gold deposits whose worth is estimated at $100 million.

Annual income from platinum dredging already in 1964 averaged about $1 million.

Off the southeast coast of Greenland, Danish enterprises are prospecting for chromite, rutile, and platinum. Near southeast Alaska, 1,000 tons of barite are mined per day. Heavy minerals have concentrated on the continental platform at depths averaging 200 to 300 meters and limenite and rutile are currently mined off the Australian coasts where 95% of the world's reserves of rutile are located on its east coast. Yearly, 450,000 tons of ilmenite are mined off the Australian west coast. Both ilmenite and rutile are often associated with zirconium and with thorium-containing monazite. Zircon and monazite are extracted simultaneously with titanium in Florida, Ceylon, and Australia. Though deposits of monazite have been located off India and Alaska, Australia remains the leader with 30% of the world's production. Near Liepaja, in the Baltic Sea, uranium has been mined since 1972. The Soviets are looking for deposits of titanium, ilmenite, and rutile in the same area. Magnetite and titanium placers have been detected in the sands of the northwest coast of the Black Sea. Similar deposits were found near Batumi and in the Sea of Azov. These placers also contain chromite – now mined off the Oregon coast – magnetite, cassiterite, and aragonite. Magnetite was exploited, until very recently, by the Finns, and the Japanese still extract 40,000 tons per year south of Kyushu Island.

3.2.6 Nickel

According to analysts, world reserves of nickel can satisfy projected cumulative needs to about the year 2000. Rapid industrialization would not cause a heavier demand because nickel use involves nonferrous alloys and special steel. The U. S. reserves are very modest, but nickel is plentiful in Canada. Nickel can also be retrieved from low grade sulphide and lateritic ores, but extraction may well be more expensive than from the sea. Oddly, manganese-rich polymetallic nodules may prove more valuable for their nickel than for the manganese because the nickel content is now higher than in land ores. (Fig. 25)

Nickel world reserves are estimated at 30 million tons and current cumulative world demand exceeds 7 million tons. Nodule exploitation could depress the U. S. manganese price by 40% and nickel price by 10%, even more if numerous producers began exploitation of this metal.

3.2.7 Sulfur

There seems to be no shortage of sulfur at this time, and, specifically in the United States, reserves are ample until after 1980. The world's demands of this chemical can be satisfied until the mid 1990s. Furthermore, there are substantial sulfur reserves in coal deposits amounting for the United States alone to some 7 billion tons, and the Clean Air Act may lead to recovery of some of it. Finally, sulfur is abundantly present in gypsum, and economically exploitable. (Fig. 26)

Prospects for exploitation of marine sulfur are good and ten years ago offshore salt domes in the Gulf of Mexico were already exploited. Sulfur exploitation rights were acquired then for $34 million on 29,140 hectares at 30 to 60 meters depth. Onshore technology was easily adapted to ocean mining. Some salt domes come close to the ocean surface and through dissolving of the halite, bacteria can bring about sulfate reduction and transform non-soluble matter into sulfur. Contemporary sulfur production from the sea accounts for more than 10% of the total sulfur production in the United States, or about 80,000 long tons in 1974. For the United States alone offshore reserves of Frash sulphur exceed 200 million long tons,* of which 100 million tons are recoverable. More sulfur is recoverable from undersea gas and oil. The term Frash designates deposits which are recoverable by drilling

---

*1 long ton = 2,240 pounds = 1.016 metric ton
1 short (U. S.) ton = 2,000 pounds = 0.9072 metric ton

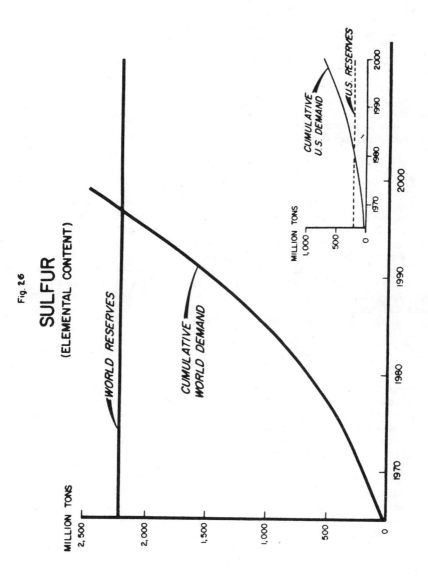

Fig. 26

SULFUR
(ELEMENTAL CONTENT)

wells into which hot water is pumped to melt the sulfur; these
deposits are found as a by-activity of the oil and gas search.
In the U.S., offshore sulphur has been exploited - mostly for consumption by fertilizer producers – at Grand Island and Caminada
Pass in Louisiana. Production reached about 2 million long tons
a year. A new market may develop as sulfur concrete competes with
cement in the construction business.

Sulfur can thus be recovered as a by-product of oil and to a
lesser extent of gas extraction.

Sulfur world reserves are estimated at 2,000 million tons and
current cumulative world demand at 450 million tons.

3.2.8 Copper, Zinc & Barite

In closing this impressive, yet far from complete, survey,
let us mention copper and zinc mining operations in Maine, potash
in Great Britain, and barite in Alaska. Here, natural forces
have not barred operations even though tides reach seven meters,
winds reach speeds of 170 km/hour and temperatures fall below
-15°C. The total production of ocean mined barite has an economic value of one million dollars yearly. Barite is also abundant
off the California coast. It is anticipated that because it is
used for the muds needed to drill oil and gas, demand could double
by the year 2000. Potash deposits occur in salt basins.

3.2.9 Muds

As a result of erosion, and from alluvial accumulations,
muds are deposited at the sea bottom at lesser depths than on the
continental platform. They can be divided into two types: calcareous muds and red muds. The first type originates from shell
deposits and can be used in the production of whitewash, as fertilizer, or for its content of calcium, potassium or barite. Such
muds cover about 35% of the ocean bottom at depths ranging from
700 to 6,000 meters with layers whose thickness is estimated at
400 meters on the average. This calcareous cover would thus
spread over 128 million square kilometers and accumulate at the
rate of one billion tons per year, or eight times the quantity
yielded by contemporary land operations.

The red muds are argillaceous and constitute a source of
aluminum, iron, copper, nickel, cobalt, and vanadium. According
to some oceanographers, they cover one-half of the floor of the
Pacific Ocean and one-fourth of the Atlantic Ocean bottom at an
average depth of 3,000 meters.

3.2.10 *Dissolved Substances*

Since time immemorial, man has extracted from sea water common salt or sodium chloride. Extraction is mentioned in Chinese writings of 2200 B.C.; it still represents some 30% of the world production today. The use of salt is both domestic and industrial. It is still extracted on the U. S. Pacific coast, in Puerto Rico, France, Sicily and elsewhere. Bromine and magnesium compounds are simultaneously extracted; 70% of the total production of bromine and 60% of the magnesium used are of marine origin. Dow, Kaiser, and Merck extract magnesium in Texas, and Dow alone provides, from marine origin, 75% of the U. S. needs in bromine. Near Los Angeles, iodine is extracted from the petroleum fields' brackish solutions.

Thick beds of tachydrite, magnesium salt ($CaCl_2$. $Mg\ Cl_2$. $12H_2O$) have been located along the Brazilian coast's Sergipe salt basin and in the Congo basin. It can probably be mined by solution methods.

The oceans represent a gigantic storehouse of 50×10^{15} tons of dissolved materials among which the chloride ion represents 54.8% of total salts, the sodium ion 30.4%, sulfate 7.5%, magnesium 37%, calcium 1.2%, potassium 1.1%, carbonate 0.3%, and bromide 0.2%. These eight ions account for over 99% of the salts. Together chloride and sodium ions account for 85.2% of the dissolved salts and are the most easily extracted.

Potash deposits in salt basins are large though not as frequent as salt and gypsum. Reserves of K_2O (potassium oxide) run into tens of billions of tons; large deposits are located under the North Sea.

Crude soda and potash is known to have been extracted from seaweed ashes in Scotland in 1720. Iodine was extracted in the nineteenth century from seaweeds, and magnesia was prepared along the Mediterranean coast. M. J. Yeaton reported in Chemical and Metallurgical Engineering in 1931 (vol. 38, pp. 638-641) that in 1923 magnesium chloride and gypsum were produced from the bitterns from solar evaporation of San Francisco Bay water. It is now produced as a by-product of petroleum and other well brines.

Commercial sea water bromine was produced from San Francisco Bay bitterns in 1926. Potassium chloride was recovered from Dead Sea waters by evaporation in 1931, and in 1932 bromine was taken from the residual liquors of the potassium plant.

A sudden surge in the demand for ethylene dibromide, a constituent of the anti-knock compound of gasoline, occurred in 1933, which could not be met by bromine from subterranean plants and led

to the construction of Kure Beach plant, North Carolina, which
five years later was producing 18,000 tons of bromine a year. A
second plant (Dow Chemical) was producing, by 1944, in Freeport,
Texas, 27,000 tons.

Dow Chemical produced the first magnesium metal from sea
water from the Gulf of Mexico in 1941, also at Freeport; within
a few years production reached 8,100 tons. The U. S. Government
completed another such plant at Velasco, Texas which produced magnesium at 30% below land-plant costs. By 1942, it was turning out
about 3,300 tons a year.

Although the Kure Beach and Velasco, Texas plants were closed
down as World War II came to an end, almost the entire magnesium
production of the U. S. is of marine origin, likewise at least
75% of the bromine. Magnesium, the lighest structural metal commercially available, one-fourth as heavy as iron and two-thirds
as heavy as aluminum, has high strength when alloyed with aluminum,
zinc and manganese, it is corrosion resistant, and easily fabric-
ated, hence very valuable in airplane and car construction. Added
prior to pouring to cast iron it gives to so-called nodular cast
iron strength and ductility comparable to steel.

While bromine production requires sea water of high and constant salinity and not organically contaminated, preferably deep,
magnesium production requires one twentieth of the water needed
for bromine extraction and temperature of the water plays very
little role. Recovery of magnesium can reach 90%. Currently the
Dow Chemical Company produces 90% of all metallic magnesium at
sea water processing plants in Freeport and Velasco in Texas.

A few years ago, hot brines were discovered in the Red Sea.
The estimated value of the mineral materials contained in these
brines is estimated at more than 1.5 million dollars. Layers,
occurring at depths of more than 2,100 meters, are sometimes 200
meters thick and contain, besides precious metals, zinc, lead,
copper, and cobalt. Joseph Lassiter, of the Massachusetts Institute of Technology, wrote that in the Atlantic Deep II of the Red
Sea, there are reserves in the upper. 11 meters layer worth $2.3
billion. Such wealth is tempting, but, so far, no extraction method has been devised. The Hughes Tool Company and Deep Sea Ventures are conducting research to develop an appropriate system.
In 1967, three separate pools of hot brines had already been defined in the Red Sea rift valley. The brines and their associated
heavy metals are believed to be periodically discharged from the
eastern side of Atlantic II, the largest deep. Seismic reflections are obtained from hot brines and make their location possible. The reflections are due to the difference in density between the hot brine and the overlying Red Sea bottom water. The
hot brine area is located in the middle of the Red Sea rift valley.

It is believed that the heat source, and some of the heavy metals, are probably associated with a local geothermal phenomenon.

The contact between hot waters and cooler oxygenated water above, brings about oxidation of the dissolved metals. Chemical and mineralogical analysis led to the division of the deposits into seven bedded and laterally correlative facies: detrital, iron-montmorillonite, goethite-amorphous, sulfide, manganosiderite, anhydrite, and manganite. Their geographical distribution, unconsolidated nature and age relations show that the chemistry of the brine has considerably changed with time. The system appears to be an ore body in the formation process. The Red Sea geothermal system is not unique, others have been located in the Salton Sea in 1963 and in Cheleken, U.S.S.R. in 1967. (Hunt, Hays, Degens, Ross, Bischoff).

Sea water composition is well known and analyses have shown quantities of dissolved copper, cobalt, zinc, gold, uranium, and deuterium. Concentrations, however, are small, yet, these elements could perhaps be recovered as a side operation of desalination plants.

If the world were to switch to an hydrogen economy, the sea would naturally be an inexhaustible source of hydrogen.

3.3   INDIRECT EXPLOITATION

Some activities are not in the strictest sense ocean mining, but they are closely linked with the ocean domain, albeit because of the technological problems involved or the location of the zones of extraction.

3.3.1 Coal *et al.*

The Japanese extract coal in mines which stretch under the sea. African industries, such as De Beers, in the Republic of South Africa, gather diamonds off the coast of the Southwest Africa Territory. In the late nineteenth century petroleum companies off California drilled through the ocean floor, first at modest depths, then at ever-increasing depths, to reach the petroleum reserves stored under the ocean. Progress in petroleum extraction in the ocean has raised questions of international law which have not yet been answered. Drilling started in the Gulf of Mexico in 1937 just prior to World War II.

Even before the Christian era, mines were under exploitation by Greeks beneath the sea at Laurium. Since then, coal, limestone, iron ore, tin, copper, and nickel have been extracted near the

British Isles, Ireland, the Atlantic coast of France, near Greece,
Turkey and Spain in the Mediterranean, Finland in the Baltic Sea,
Japan and China in Asia, and on the American west coast off Alaska,
Canada, the United States, and Chile.

Coal, like gas and petroleum, is not extracted from the ocean
but rather from under the ocean. Such coal mines are operated in
China, Japan, Turkey, Great Britain, Nova Scotia, Newfoundland, and
Chile. Some two million tons were extracted in 1965. While petroleum and gas production is progressing at an accelerating pace
in the North Sea, important reserves of methane have been discovered and are also being tapped. A listing of indirect exploitation
of the ocean should include such products as corals, used to manufacture jewelry (Bahamas and Hawaii, amber, Baltic), and lime-
stones, used for building and cement manufacture.

Among the organic geological resources are shells, petroleum,
and natural gas. We mentioned shells earlier: as for gas and
petroleum, they are being depleted at a faster rate than extraction progresses. Not so long ago, wells dug at depths of 25 meters were considered quite an achievement; today depths of 250
meters are not unusual and current prospecting is considering deposits more than 400 meters under the water surface. Petroleum
demand doubles approximately every ten years, a situation which
has stimulated worldwide prospecting.

3.3.2 Hydrocarbons

Hydrocarbons, however, remain the most abundant mineral resource extracted from the oceans today. One of the richest fields
underlies the North Sea. Some finds are spectacular because of
their reserves, others because of the technological difficulties
in tapping them. In 1972, one find was sited 135 miles off Aber-
deen, Scotland, with an estimated daily production of 4,000 barrels of petroleum and 60,000 cubic meters of natural gas. The
North Sea "Brent concession" is about 100 miles north of the
Shetland Islands, at 91°N, Latitude, at a depth of 150 meters, in
an area where winds reach speeds of 160km/hour and waves are frequently 30 meters high. The Soviets have found considerable gas
and petroleum reserves east of the Crimean Peninsula in the Black
Sea, along the Romanian Dobrudja littoral and on the submarine
plateau stretching from Varna, Bulgaria to the Bosporus. Monthly
petroleum industry periodicals list new sites, new areas under exploration, new concessions being granted. But this flurry of economic exploitation of the oceans is not without dangers. Petroleum
companies and shipping companies are severely taken to task by
environmentalists. Legal attempts in 1977 and 1978 to block oil
drilling on George's Bank and in the Baltimore Canyon have proved

partially successful.

3.3.2.1 Oil and Environment

After disastrous experiences encountered with rigs and spills, whether major (over 4000 barrels), moderate (240 - 2000 barrels). or minor (less than 240 barrels), the oil industry has kept a reasonable record. In 1970, for instance, 3 fires and explosions occurred in the Gulf of Mexico which took weeks to burn out, 1 gas explosion occurred in Australia, another in the Persian Gulf, still another in Indonesia. During the same year, the biggest explosion shook offshore Louisiana and it took four months to put out the fire. On the other hand, only one major fire broke out and that was near the Iranian coast in 1971. Further efforts at prevention of environmental damage by oil companies has helped to improve a badly tarnished image.

Collisions of giant tankers, wells out of control, blowout of a layer, or fires on drilling platforms, all contribute to disasters known as the black tide, with its ensuing pollution of water and air, destruction of aquatic life, and spoiling of ocean beaches. Though cases brought to court are dealt with severely, unbridled commercial efforts continue. The need is such that there are not sufficient drills to satisfy the demand, especially those for great depths. In 1971 one company brought a giant drill back from Australia to the North Sea at a cost of two million dollars and found the investment profitable.

A mammoth self-elevating oil drilling rig has been developed for offshore oil exploration in the water of the North Sea and was constructed in Clydebank, Scotland. It withstands rugged weather, freezing temperatures and the battering of 20 meter waves common to the United Kingdom's waters.

From the bottom of the 118 meter supporting legs to the uppermost derrick, the rig measures more than 1,525 meters. Electrohydraulic jacking assemblies adjust the legs of the rig to operate in waters up to 84 meters deep. Twelve hydraulic jacks lift the rig's platform. Rubber shock absorbers are positioned at the top and bottom of each jack. They are engineered to take a work load of up to 300 tons with a minimum deflection of 3.8 centimeters.

As of 1969, one fifth of the world's petroleum production was of submarine origin and close to 400 million tons of gas and oil had already been extracted from the ocean milieu. It is foreseen that by 1980, or 1985, wells will be drilled at depths of 2,000 meters. Considering that, in 1965, 15% of the world production of liquid hydrocarbons came from the ocean, in addition to 6%

of the gaseous hydrocarbons, the increase is about 5% per year.
It is foreseen that by 1980 30% will come from the ocean, and by
1990 more than two billion tons will be extracted, which represents the total world production of liquid hydrocarbons in 1971.
Within 20 years, the number of drilling platforms grew from none
to 360. Yet, many believe that these quantities will not satisfy
the demand.

From a technological viewpoint, oil drilling operations still
follow the traditional pattern involving a derrick and a drilling
platform, some of which are sufficiently large to accommodate a
heliport. Humble Oil is pursuing research to refine a remote flow
control system; diving bells would be used to periodically inspect
installations, while the oil itself would be stored in pre-stressed
concrete tanks resting upon the ocean bottom.

Ultra-modern oil storage facilities in the North Sea, coupled
with successful experiments conducted with pre-stressed concrete
tanks on the bottom of the Red Sea, show the continuing efforts
to minimize the environmental and financial costs of oil spills.
Underwater storage capacity in Dubai reached 10^6 tons, and
200,000 deadweight tons ships can now load there.

3.3.2.2 Demand, Production and Reserves

Oil demand estimated at 8,000 million tons by 1980 and the
need to spread the geographical sites of operations so as not to
be at the mercy of political crises, are sufficient motivation to
drill at ever greater depths even though the cost of installations
doubles each time depth increases by 600 meters. Since 1970, 75
countries have engaged in geological and geophysical exploration
along their coastlines and 70 have drilled in the sea, with production occurring in at least twenty nations.

Offshore crude oil reserves are estimated at between 100,000
and 120,000 billion barrels, or 20% of the world total, and 14
million cubic meters of natural gas. In the single year 1972,
the world offshore production equalled roughly 9,000 million
barrels of oil (U. S. alone 5,500) and 864,000 million cubic meters of gas (U. S. alone 680,000). According to recent U. S. Geological Survey data, the U.S.A. has left from 70 to 90 billion
barrels of oil and between 11.3 and 14.2 trillion cubic meters of
natural gas of which respectively 10 to 14 billion barrels and
2 to 26 trillion cubic meters are offshore in waters not exceeding
200 meters in depth. Drilling rigs dot the Gulf of Mexico, 240
off Louisiana and Texas as far 95 kilometers from shore where the
bulk of U S. reserves lies, and the North Sea at depths of over
150 meters and 160 kilometers from shore. In California's Santa
Barbara channel, drilling has gone beyond 300 meters depth. The

ocean off Alaska is considered as at least an equal prospect as
the Gulf of Mexico. Four hundred rigs are in current operation
worldwide and half are mobile. Drilling has exceeded in some sites
500 meters depth, and offshore distances of 265 kilometers reached.
New records are foreseen as much as to 4,000 meters. Current offshore drilling accounts for 8 million barrels a day, which represents about 17% of the world's total of 48 million barrels. Over
15,000 offshore wells have been drilled in U. S. waters. Would it
become technologically feasible to drill at depths of up to 2,500
meters, U. S. reserves would nearly double. (Table 9)

It is estimated that during the 1970s alone, world oil consumption will have topped 235 billion barrels, or more than was
used in all the years prior to 1970. The total world consumption
of 18 billion barrels in 1971, was about 1/35th of total estimated
reserves. (Fig. 27)

With the demand for petroleum perhaps as much as 8 billion
tons by 1980, drilling takes place at ever greater depths.

New oil finds, besides those already mentioned include: the
North Sea with production climbing to reach 100,000 barrels a day
from the British Beryl field alone; the Gulf of Suez where at
3000 meters depth an oil bearing layer 70 meters thick was located;
Nigeria, with a potential flow of 8,800 barrels a day in five
separate areas some six kilometers off the coast; along the coasts
of Canada (Sable Island) and at several sites off the coast of
Louisiana. Between Sharjah and Iran a new field off Abu Musa Island, will, if political events permit, start production soon.

Meanwhile, further search operations, not all successful of
course, were carried in Far East, North Africa and Australian
waters. Renewed Soviet production is foreseen in the Caspian
Sea's eastern continental shelf.

Ekofisk, the world's largest offshore one million barrel oil
storage tank was towed 330 kilometers from Stavanger Fjord into
the Norwegian sector of the North Sea. Seventy meters of the
total 95 meters' height is under sea level. Now an "artificial
island", it is about 2 meters above the sea floor. The in-between
space will be filled with mortar. (Fig. 28) Ekofisk One is
"anchored" in the richest oil field of the North Sea; the Ekofisk
field first producing 45,000 barrels, is now yielding 450,000
barrels per day in 1978, or about 25,000 tons annually, and 12,000
million cubic meters of natural gas per year, as well. Meanwhile,
the considerable energy required to operate this installation is
provided by the gas from the field itself.

The United Kingdom expects to raise its North Sea oil production to 580 million barrels per day by 1980. Another strike in

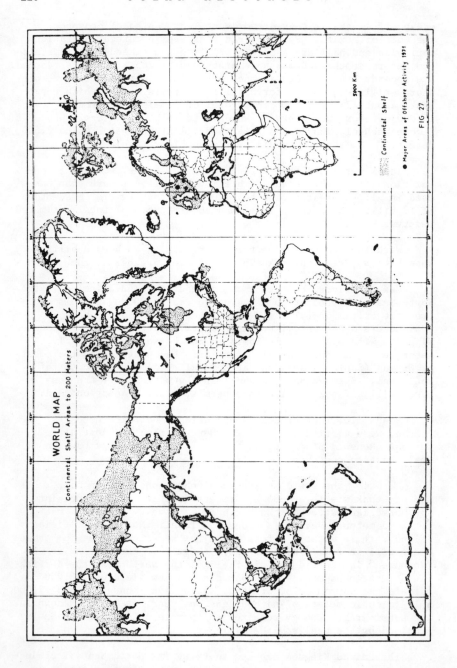

WORLD MAP
Continental Shelf Areas to 200 Meters

Continental Shelf
● Major Areas of Offshore Activity 1971

5000 Km

FIG 27

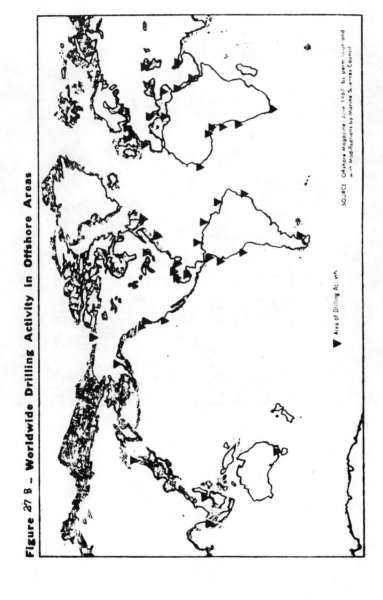

Figure 27 B – Worldwide Drilling Activity in Offshore Areas

▼ Area of Drilling Activity

SOURCE: Offshore Magazine June 1967 by permission and with Modifications by Marine Sciences Council

Fig. 28   Ekofisk Platform in the North Sea.

the Brent field off the Shetland Islands, may yield 4,000 barrels
per day. A new huge strike near Scotland and still two more
strikes were recently reported close by. In only eight years,
counting only United States federal waters, the North Sea has
surpassed the Gulf of Mexico production.

Current oil consumption outside of the Communist areas was
about 12 billion barrels annually, and at this rate it may reach
27 billion by 1980. However, most experts agree that future discoveries will not keep pace with demands nor needs, this regard-
less of the fact that offshore drilling alone will provide by 1985
and not an additional estimated 1 1/2 billion barrels annually. Oil
use increased spectacularly over ten years. In 1963, world consumption was, in thousands of barrels per day, 18,290, but in 1973
it reached 35,040, an increase of 92%. Of this amount, 10,550 and
15,980 were, respectively, in 1963 and 1973, consumed by the United
States alone, a solid 51% increase, but falling far short of the
spectacular Japanese increase of 284% when that country jumped
from a consumption of 1,250 barrels in 1963 to 4,850 in 1973.

With consumption still outpacing discoveries, a shortage will
develop in the Western world notwithstanding the new reserves
added to the known ones. The threatening shortage is well evi-
denced when world consumption of 1963 is compared to that of the
present.

Oil and gas now account for 65% of total world energy consumption. Since 1974, petroleum demand has risen at an annual rate
of 7% to meet it. This means a production capacity of 4 million
barrels a day must be added, or 20 billion barrels a year. By
1985, the daily needs would reach 225 million barrels. Between
1960 and 1970 an annual rate of discoveries, outside Communist
countries, of 19 billion barrels a year was reported, but consumption rose from 9 to 17 billion barrels a year! New energy production, for the immediate future, and exclusive of the harnessing
of ocean water energy, could come, states an Exxon report, if we
delay fast nuclear development, respect environmental impact rules
on coal use and on transportation of natural gas, from a 50% production increase of Middle East oil, 18% in other   oil, 9% in
natural gas and in coal each, 10% in nuclear power, and 4% in
hydroelectric power.

According to the Oil and Gas Journal (December 25, 1972),
the world known oil reserves are distributed among the Middle
East, in whose hands the key rests – predominantly with Saudi
Arabia (53%), Africa including the "Arab" countries (16%), the
Soviet block (15%), the United States including Alaska (5%),
America south of the U. S. border including the Caribbean (5%),
and last Canada, Europe and Indonesia (each 2%). New sites are
even searched for in Arctic and Antarctic waters and new techniques

could reduce costs substantially (1972, _World Oil_, Vol. 175, No.5,
pp. 71-72); so can probably offshore testing (1972, _World Oil_,
Vol. 175, No. 6, pp. 48-51). The Glomar Challenger expedition of
1972-73 revealed that traces of natural gas were found off Antarctica beneath the sea floor. Mexico in the past decade has also
discovered large reserves of oil. (Table 9)

3.3.2.3 Current Search for New Source Areas

The search for oil is especially directed towards Southeast
Asia where considerable investments for research are being made.
According to figures released in June 1974 by the Chase Manhattan
Bank, the oil industry's investment will amount to $55 billion in
Asia, and $23 billion in the Mid East for the period 1970 - 1985.
Yet, the biggest producer in Asia, Indonesia, pumps only 1.1 million barrels a day, while Kuwait exceeds 3 million, Iran 5 million
and Saudi Arabia 8 million. However, with price increases of oil
surpassing 300% in 1974 alone, even small fields are economically
worthwhile. Already over 40 rigs are operating in Asian waters,
mostly Southeast Asia, and 100 oil and oil related firms, American, French, Italian, West German, British and Australian, have
established offices in Singapore. The reasons for the massive
investments in that part of the world, include (1) Middle-East
fields are known, but Southeast Asian fields are not; (2) a desire to attain a certain independence from the Arabian producers;
(3) Southeast Asian oil contains less sulfur; and (4) technological progress makes drilling easier and possible at greater depths.

3.3.2.4 Natural Gas

Because demand for natural gas is steadily increasing, plans
must be made to transport it over relatively long distances and
in an easily transportable form. Conversion to liquified natural
gas is such a form, but construction of liquefaction plants runs
between $500 and $700 million, hence fields must be either large
or several smaller fields must cluster closely. Pipeline transportation is possible for fields located in the North Sea, Gulf
of Mexico, and Northern Canada. Pipelines such as those used in
the North Sea and the Gulf of Mexico provide good transportation.

The largest reserves of gas are found in the Soviet Union
from the Pechora basin to the Dnieper-Don basin and throughout
Soviet Asia. On the West Coast of North America, gas is available
in Cook Inlet and Bristol Bay; South America has potentials in the
Gulf of Guayaquil and likewise in the Caribbean in the vicinity
of Trinidad. Africa has rich reserves in Algeria, Libya and the
Niger Delta. In Asia, the Persian Gulf retains a leading position
(Iran, Abu Dhabi, Bahrain), but great progress is made in Brunei,

Sarawak and Indonesia. Offshore fields have been located on the
Australian Northwest Shelf and in the Cook Straits separating New
Zealand's North and South islands. Southeast Asia has reserves
in Vietnamese waters and in the Gulf of Thailand. King Potential
LNG sources: Ocean Industry VIII, 11, 35-38, 1973, estimates gas
reserves beyond those of Europe, Canada and the United States at
about 23.6 trillion m^3 (834 trillion cubic feet) not counting the
Gulf of Thailand and the South China Sea basins. In this total
the Persian Gulf and the U.S.S.R. account alone for respectively
8.5 trillion m^3 (300 trillion cubic feet) and 6.3 trillion cubic
meters (220 trillion cubic feet). An additional 0.14 trillion
cubic meters (5 trillion cubic feet) is expected by 1985 from
offshore drilling alone. Consumption of liquid natural gas has
also climbed spectacularly. In 1973, the United States, Japanese
and European markets grew rapidly. (Table 5)

Table 5.   World Natural Gas Market Consumption (1973)

MARKET	FROM		OUT	
	in m^3 (millions)	in cu. ft. (millions)	in m^3 (millions)	in cu. ft. (millions)
United States	127	4,500	323	11,400
Japanese	89.2	3,150	174.2	6,150
European (actual)	59.5	3,100	119	4,200
(actual & planned)	92	3,250	151.5	5,350

The United Kingdom alone with natural gas reserves estimated at
0.9 trillion cubic meters (32 trillion cubic feet) expects to
raise its North Sea fields production to 0.1 billion cubic meters
(4 billion cubic feet) a day. Canada's reserves are calculated at
1.5 trillion cubic meters (55 trillion cubic feet).

3.4 CONCLUSION

The potential mineral wealth of continental rocks is enormous and submerged continental areas are large enough to contain
huge quantities of minerals which eventually will make important
contributions to world economy. Placer deposits are richer and
more extensive in some regions, such as in southeast Asia. Qual-
itatively, if not quantitatively, submerged parts of continents
add substantially to land reserves. The mineral resource potential of the deep ocean is small per unit area as compared to that
of continents because there is no thick sialic crust, the seat of
ore-producing magmas, no rocks pre-dating the Cretaceous are known

to be exposed, and important sedimentary rocks such as coal, evap-
orites, bedded-phosphate are not known to have formed in the ocean.

Ocean mining may have serious socioeconomic consequences if
it displaces land mining; places and people may be affected. Relationships of marine minerals to onshore minerals must be taken
into account, and we must aim for economically efficient operations.
If marine mining presents less waste disposal problems than on
land operations, then this could be an incentive to turn to the
sea, but our knowledge is still limited.

Table 6

Petroleum Demand
(in millions barrels per day)

GEOGRAPHIC AREA	YEARS					
	1960	1965	1970	1975	1980	1985
U.S.A.	20	23	25	30	35	40
Europe	10	11	13	20	25	35
Far East	2	3	4	7	10	15
Other Non-Communist Countries	1	1	2	3	6	10
Communist States	2	3	10	15	20	40
TOTAL	35	41	59	75	96	140

Table 7

Growth of Oil Discoveries
(in billions barrels/year)
By Year Anticipated to 1990

YEARS

1920	1930	1940	1950	1960	1970	1980	1990
	8		15	15	22	18	21	22

Table 8

Comparison of Land and Offshore Wells Throughout the World

YEARS	ONLAND	OFFSHORE
1960	600	105
1961	750	106
1962	900	110
1963	1050	120
1964	1200	125
1965	1600	160
1966	1800	220
1967	2100	280
1968	2300	315
1969	2200	360
1970	1900	400
1971	1850	410

Table 9

World's Estimated Oil Reserves (in billions of barrels) (1972)

GEOGRAPHIC AREA	AMOUNT	% (APPROXIMATE)	*
Middle East	367.4	58.5	(53 *)
Communist world	98.5	15	
Africa	58.9	9	(16 *)
U.S.A. (incl. Alaska)	37.3	6	(5 *)
Central & South America	31.6	5	
Asia & Pacific area	15.6	2.5	
Europe	14.2	2	
Canada	8.4	1	(2 *)
TOTAL	631.9		

* Exxon's estimates

III

THE OCEAN AS REAL ESTATE

OCEANIC bottoms represent very important areas of real estate
today. The Dutch viewed it for centuries as territory to be recovered, dammed, drained and transformed into agricultural land.
They have become masters at the art of land reclamation, and added
"polderland" to their territory for over five centuries. (Fig 29)
The major project, started in 1928, has reduced a deep sea in-
dentation in The Netherlands known as the Southern Sea (Zuiderzee)
to a mere lake, Yssel Meer, bringing about deep hydrological,
economic and social changes for the northern regions of the country. (Fig. 30) Spurred on by the disastrous floods that raged in
1953, insular Zeeland and the neighbouring regions are now com-
pleting the ambitious Delta-Plan, which foresees closing the
Eastern Scheldt and linking the Scheldt delta islands by drained
land. This will shorten the coastline, reduce flood threats, and
create a fresh water basin. Rotterdam and Antwerp will retain
free access to the sea via the New Waterway and the Western
Scheldt. Whether this infringment of man on the sea will not
cause hydrological problems remains to be seen. (Fig. 31)

Without modifying coastlines, the ocean remains very valuable
as a site for pipelines, transmission lines, cables, recreation
resorts, underwater cities, artificial islands, and transport
lanes.

1. CABLES AND PIPELINES

For more than a hundred years telephone and telegraph cables
(Italy-Sicily, U.S.A.-Liverpool) have rested on the sea floor.
The last such cable to link Japan and Hawaii was completed in
1964. Other recent cables linked Naoetsu (Japan) and Nakhodka
(U.S.S.R.), Newfoundland-Greenland-Iceland-Scotland, Hong-Kong
and Malaysia.

The first transoceanic (distinct from transmarine) cable
was inaugurated in 1866. In 1956, the first TAT-1 cable was installed and therewith transatlantic voice-grade connection with
Europe became possible. These two dates are milestones in the

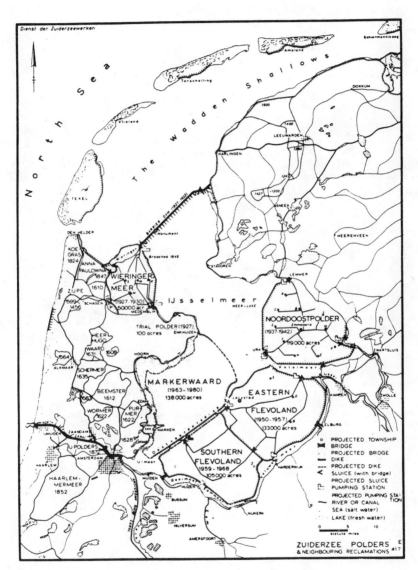

Fig. 29 Reclamations in The Netherlands

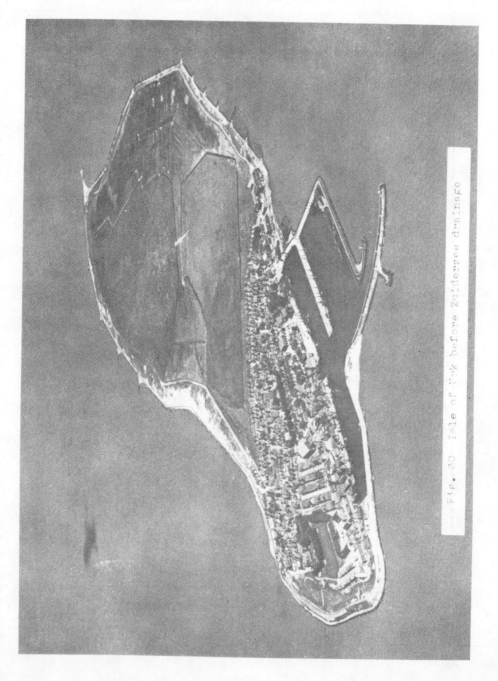

Fig. 40 Isle of Urk before Zuiderzee drainage

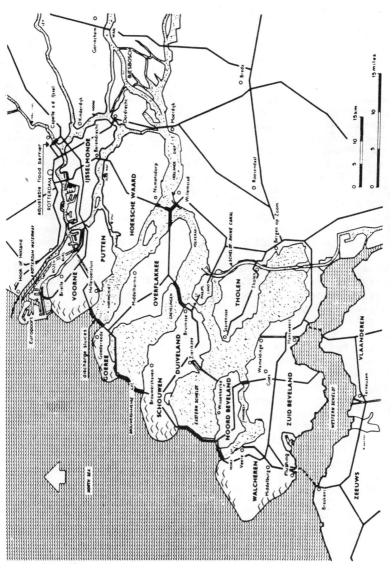

Fig. 31 Map of The Netherlands Delta-Plan

history of communications: the first one signals the divorce of
transportation and communication; the second heralds a progressive
regression of the importance of cables. Satellite technology is
rapidly providing new ways of communications. Satellite communication facilities are leased by COMSAT (Communications Satellite
Corporation) to the traditional carriers such as A.T. & T., R.C.A.,
I.T.T., Western Union International, and others.

According to a 1974 report to the U. S. Senate Committee on
Commerce, the future of ocean-based communications systems will
be governed by three variables: first, the growth in total volume
of international communications; second, the growth in revenues
and the share attributable to transoceanic transmission; and third,
the transoceanic share of said revenues. This last element is
the one we are most closely concerned with. Doubtlessly satellite
transmission is less costly than cable communication and is technically more flexible. It might be advisable to keep both systems
operating just in case of a failure of one. Other considerations
favoring the continuation of cable service include national security and political and economic relationships with European gov-
ernments which might want to protect their cable manufacturers.

The substantial economic rent derived from underwater cable
activity has declined considerably. It is projected that over
the last quarter of the twentieth century the share of cable
transoceanic communications will drop from 48% to around 25%.
While this the use of the ocean floor might gradually decline as
a result in the changing nature of means of communications, it is
increasingly important for oil and gas transportation. Pipelines
(oleoducts) link production fields to loading harbors, and gas
pipelines (gasoducts), where practical, are a very economic alternative to natural gas liquefaction. Many a promising location
for an electricity generating plant is left unused because of
distance to consumer. In this instance too the ocean can play
an important role since conventionally generated electricity is
also being transmitted by undersea cables, permitting grid to
grid transfer.

2. TOURISM AND RECREATION

Underwater tourism, once a science fiction dream, was a
highly popular attraction during a world exhibit in Geneva, when
Piccard used his mesoscaphe to take tourists on an underwater
ride in Lake Geneva. In Florida and the Virgin Islands, the
U. S. National Park Service created, several years ago, the two
first underwater national parks, providing scuba divers with an
opportunity to observe, along marked paths, aquatic flora and
fauna in an undisturbed environment. Popular response was such
that a third such park was inaugurated in Hawaii in 1970.

The coast itself is severely taxed by recreational use. For
instance, in 1945, New York public beaches accommodated 5 million
visitors a year, but in the early 1970s the total had reached 61
millions. Very little effort has been made to protect beach and
dune or the coastal waters. Recently, at international, national
and regional conferences, panels addressed themselves to such
aspects as: trade-offs between natural areas and development;
compatibilities among recreation users for multiple use; extractive versus non-extractive use of living resources; restoration
and management of natural zones; the conflict between tourism and
environmental protection; public access to recreational sites
(6.5% only of the entire U. S. Atlantic-Gulf coastline is publicly owned); and evaluation of recreational resources. The "marine
recreational potential" includes estuaries, coastlines, nearshore
waters, and coastal islands. It involves rest, wildlife study
as well as high intensity uses such as boating and sportfishing.

Recreational use of the coast may become less viable because
of the higher returns of industrial use. Such conflict is even
the more apparent in densely occupied areas, as seen by the Belgian coast where tourism and leisure time use has run afoul of
major industrial projects such as construction on land and offshore of the North Sea port near Zeebrugge. (Noordzeepoort)

The choice of coastal area use is hard to make since the
democratization of travel and the increase of leisure time after
World War I turned tourism into the major source of revenue for
some countries. Spain, Israel, the Communist nations earn much
of their "hard" currency from tourism. The sea has always been
a major magnet, but exploitation of its recreational potential has
raised conflicting demands. Some commercial interests have been
directed towards the consumer and providing services, but this
seasonal migration of the population towards the shores competes
with industry's search for water, energy and other resources.

Tourism now accessible to the popular masses has given rise
to an ever-expanding industry. Water tourism attracts millions
of men, women, and children, and ocean shore recreation is perhaps the leader in this type of activity, including fishing, nautical sports, swimming, and just plain beach relaxation. According
to a National Recreation Survey made in 1960, the average U. S.
citizen between ages 12 and 75, spent about 15 days or 260 hours
in such activities, and by 1972, these totals had reached 17 days,
or 300 hours. The estimations for 1981 were 19 days or 345 hours,
an average annual growth of respectively 1.35 and 0.7%. Strictly
ocean related, therefore not including camping, hiking, bicycling,
and other land-based recreation, are 22% of these recreational
and tourism activities. Unfortunately, this development caused
problems in the development of shore resorts. Promoters fill
estuarian waters to create so-called shoreline property and

endanger, or fully destroy, the spawning grounds of many fish and
shellfish. Beach erosion, dune bulldozing, lack of sanitary facilities, damages caused by the tourists themselves, and commercial developments, have wrought havoc with some coastal areas,
brought about the virtual disappearance of entire stretches of
beach, and in some instances has led to the toppling of valuable
properties and buildings along the shore. Unless we have given
up hope of salvaging the tourist resources constituted by the
ocean, control measures must be taken immediately. The United
States has already passed strict legislation to protect the shore
areas, but only a few other Western countries have managed to
protect their natural sites.

Some slight slowing down of the coastal zone's tourist development may occur due to inordinate crowding, cost of travel,
beach erosion, and improvement of environmental quality of inland
bays and rivers. However, it is more likely that competing demands upon a necessarily limited coastal area by different "users",
will cause a further escalation of the problems. Allocation of
space becomes a necessity, and recreation is only one of these
"users". In the United States, the Coastal Zone Management Act
of 1972 is aimed at encouraging planning but very little has been
accomplished thus far. Progress is a slow movement in this
direction.

3. ARTIFICIAL ISLANDS AND UNDERWATER CITIES

The futuristic views of Athelstan Spilhaus, former dean of
the Institute of Technology at the University of Minnesota, were
considered too imaginative a decade ago. Even though underwater
cities and resorts have not left the drawing boards, the Japanese,
surrounded by an incredibly polluted environment, plan to build
housing complexes beneath the waters of the Sea of Japan. If
construction has been delayed, it is not because of technological
handicaps, but because authorities want to ascertain that fishing
and the coastal environment will not suffer ill effects from such
a project.

At Oceanexpo-1975 on Okinawa, the Japanese displayed a re-
markable floating city, Aquapolis, which anchored in a bay, could
be raised or lowered according to sea conditions. Some objections
were raised because it negatively influenced an existing coral
reef. Artificial islands and bottom settlements can be separated
into four major types: (1) floating cities - these could be mobile, pulled by tugs to shore to bring workers to land, then returned "to sea"; in fact such constructions are more ship than
island. Buckminster Fuller, describes floating triangular artificial "atolls" in his book "Floating Cities" (World, 1972);
(2) marine cities - these include Japan's Kikutake complex city

of linked surface and submarine units - and Paolo Soleri's floating cities Novanoah I and II described in "The City and the Image
of Man" (Massachusetts Institute of Technology Press, 1968); (3)
anchored cities - in this category we may consider Aquapolis.
Such cities could be built and assembled on land, then towed to
their location at sea and anchored permanently. In this class
falls also the modern stilt city, the thalassopolis of Jacques
Rougerie, designed to relieve, offshore, the population jam of
Indonesia; and (4) bridge cities - resembling Venice, these cities
would be built by sinking heavy caissons made of steel and concrete. The built-in pillars would be modern stilts upon which
platforms would rest linked to one another by bridges.

"Ocean Venices" could accommodate 30,000 people. Bos Kalis'
proposal for the Westminster Dredging Company would construct in
the North Sea an artificial archipelago whose islands would have
specific purposes: 50-hectare surfaces for domestic and industrial wastes treatment plants, 300 hectares units to accommodate
airports and industries, and 1,000 hectares islands to house power
and desalination plants. Power for these islands would be the
natural gas tapped, in situ, from beneath the ocean bottom. Natural gas discoveries are an incentive for building such islands.
But so is the ever increasing size of ships unable to manage coast
and inland ports. Exploitation islands, e.g. for aragonite, already exist at Ocean Bay in the Bahamas. Another example is
Termisa, off Brazil's State of Rio Grande del Norte.

4. THE OCEAN AS A HIGHWAY

The ocean offers travel lanes in depth and in surface, un-
hampered so far, and its shores provide choice locations for the
establishment of harbors. The intensity of today's traffic on
some routes and towards specific ports. and the ever increasing
tonnage of ships have caused traffic jams, collision incidents,
and far too many accidents which often result in the spilling of
dangerous, dirtying, and noxious substances. It is high time
that some order be brought to the maritime lanes, for safety, for
environmental protection, and that laws be updated.

On the one hand ship sizes have become awesome, consider the
484,000 ton deadweight tanker built by Japanese yards for British
Globtik, and a 500,000 ton ship for Stratis Andreadis. The number is increasing - 31,520,373 tons were launched in 1973 - and
on the other hand, world seaborne traffic has more than doubled
over the last 15 years. More than 750 ships pass each day through
the straits of Dover.

Transport of petroleum depends heavily upon ocean routes, and
tankers account for 66% of world trade and 50% of world non-mili-

tary tonnage. In 1971 alone 1300 million tons of oil went by sea.
While the largest tanker in 1948 had 26,000 deadweight tons, in
1974, 400 tankers exceeded 200,000 deadweight tons. Liquid natural gas carriers are also increasing in average capacity from
some 58,000 m^3 in the past decade to a projected 130,000 m^3 in
the next.

Increasing size and speed of energy carriers raises new problems with regard to safe transit through straits, and requires
improvements in navigational accuracy in sea lanes and near coasts,
charts, and standards of professional training for seafarers. In
addition, the trend towards vessels with draughts in excess of
20m severely limits the number of ports available for loading, un-
loading and distribution of petroleum and makes navigation dangerous in shallow seas. This is leading to the development of offshore tanker terminals (which in the future may be combined with LNG
carrier terminals and petroleum processing) and substantial offshore petroleum storage. Large seabed storage tanks have been
installed in the Persian Gulf and in the North Sea and many more
are planned. Complex systems of undersea pipelines link offshore
oil and gas fields and tanker terminals with means for distribution ashore.

Even though passenger traffic by sea is all but a memory, the
sea surface is thus criss-crossed by huge vessels whose sizes have
caused accidents and required new harbors to be built.

Obviously, the occupation of the ocean for exploitation has
begun, while ever more ships travel its lanes.

IV

THE OCEAN, IDEAL DUMPING GROUND

ONE OF THE strangest uses the ocean is subjected to is that of a
major receptacle of wastes. Unfortunately the mushrooming world
population is now overtaxing the power of the ocean to relieve
man of his "garbage" and still regenerating itself. Either directly or through the channels of rivers, industrial and domestic
wastes end up, often totally unfiltered, in the oceans. The
ocean, in turn, can decompose these wastes, with the help of bacteria and the sun. This power, however, is limited. Jacques-
Yves Cousteau once said that due to pollution and overfishing, he
estimated the ocean to be 40% destroyed. Over 35 years of observation led him to the conclusion that coral reefs have been nearly
halved, and some fish are becoming; endangered benthic life, as
well, said he, has been greatly disturbed.

William Bascom, however, vehemently disagrees with Cousteau's
diagnosis of a polluted ocean and maintains that the ocean can
stand more dumping. Yet Spanish researchers report that marine
life in their waters was reduced 40% over 20 years. One thing
remains certain, garbage disposal along the shoreline is disgrace-
ful and some shore waters are frequently unfit for bathing and
barely favorable for marine life if at all. Changes in the estua-
rine environment brought on by landfills, dredging, damming, insect control, and pollution, may well be damaging to species
stocks that are sedentary, but also to those fish that are sought
in international waters. Such pollution problems lead to internal
and international jurisdictional conflicts and point to both the
need for legislation and criteria by which to judge the relative
values of conflicting uses.

As Christy wrote: "Questions about the use, protection and
distribution of the sea's wealth arise because of the vacuum created by the absence of national jurisdiction, together with the
growing economic importance of the resources. But conflicts are
also occurring because of congestion in the use of ocean space
... there are too many users of the same resource, a manifestation
of economic inefficiency ... [and] different users [wish] to occu-
py the same space for different purposes. For example, oil drilling is conducive to oil spills that contaminate large ocean areas

133

and oil rigs constitute a navigational hazard even though they are
not always placed in the choice extracting site.

1. POLLUTION IN THE MEDITERRANEAN

Signed in 1954, the Oil Pollution Convention was not enacted
until 1967, and the crucial 1969 amendment closing two Mediterranean areas where oil may be discharged, is still not enforced for
lack of a dozen additional signatories. Only Australia and Japan
signed the tough 1973 anti-pollution, no oil-discharge, Mediterranean convention! Yet it would cost less than $100 million to pro-
vide harbors and refineries with the ballast cleaning equipment
needed for the tankers.

While the actual quantity of waste materials actually reaching the oceans is very difficult to assess, coming from rivers,
bays, atmosphere and human action at sea, solids, liquids and
gases, it is certainly quite large. B. H. Ketchum who edited
"The Water's Edge: Critical Problems of the Coastal Zone" (Cam-
bridge, Massachusetts Institute of Technology Press, 1972) estimates that 2 million tons (metric) of zinc, copper, lead, and some
other heavy metals reach the ocean each year, to which we must add
at least 93 million tons of hydrocarbons including 2 1/2 million
from vessels and industry, and between 20 and 200 million tons of
nitrogen rich nutrients.

During one year, dumping off the U. S. coastline consisted of
52 million tons of dredge spoils and 10 million tons of industrial
wastes and municipal sludge.

Ketchum underscores that data is scarce on the environmental
effects of ocean pollution, effects which are a function of the
nature, the quantity and frequence, the spatial concentration, and
water and ecological characteristics. In the controversy of the
bell ringers and the sceptics, the latter consider that the ocean
can and does assimilate most quantities of wastes without harmful
results if such wastes are not accumulated in shallow areas where
flow and mixing are insufficient.

The economic "rent" of the ocean as a dump is extremely high
and would still increase were it not for the concern of pollution.

There are no studies currently completed which could be used
as standards of tolerable limits of ocean dumping, and polluting,
nor is there a consensus as to the choice to be made between "open
sea" and "shallow waters". These two positions have been discussed
by D. W. Hood et al. and Jannash and Wirson, respectively.

Tourism is a victim of ocean pollution. Bathers do not

return to beaches where black tarry oil sticks to bathing suits
or feet and shoes. Beirut's sewer outlets pour 55 million tons
of untreated sewage where water skiers skim the surface oblivious
to its threat; Comino's Blue Lagoon in Malta is surrounded by an
oil stained beach; St. Tropez is spoiled by oil residue; half of
Genoa's beaches are affected by sewage pollution; Berre Lagoon
(Camargue) and Muggia Bay are virtual biological deserts as a result of industrial pollution.

Besides tar and oil, the Mediterranean Sea is polluted by
sewage, 90% of which is untreated. France, Malta, Israel are
taking measures. The third pollution source, industrial wastes,
endangers the Spanish, French and Italian rivieras, and the
beaches of Israel and Lebanon.

2. BEACH POLLUTION

Beaches are polluted in Yugoslavia, the North Sea, the English Channel, France, the Baltic Sea, in Brazil, along the coasts
of the Gulf of Mexico, and the Atlantic and Pacific United States.
Every quart of Caspian Sea water has passed through an industrial
plant and has hardly been purified. Until 1965, raw wastes and
used oil were dumped into it. William Brumfitt of Royal Free Hospital, London, estimated that in 1972, a Mediterranean Sea bather
had a 15% chance of becoming infected with a viral disease. A
few years before the French periodical "Que choisir?" published a
list of endangered beaches stretching from Norway to the Mediterranean shores.

3. SLUDGE, NUCLEAR AND DREDGED MATERIALS

As mentioned previously, the ocean has been used as a garbage
dump, but this was mostly for organic matter, which to a point
may be considered a fertilizer. However, the pouring of inorganic
compounds and poisons constitutes a different menace. Washing
out oil tankers at sea and discharging oily ballast from fuel
tanks of cargo vessels inhibits lobster, crab, shrimp breeding,
spoils fish, ruins clams, kills birds and may change migration
routes of fish, leading them away from their breeding ground.
Furthermore, sea organisms selectively fix wastes and poisons.
Fifty parts per million of D.D.T. has been found in fish 80 kilometers offshore.

Leakage from concrete covered steel drums filled with radioactive wastes dumped into the sea constitutes a great danger
even if sunk into deep still water or fast sedimentation zones.
Use of atomic explosives to dig channels entails unknown risks.

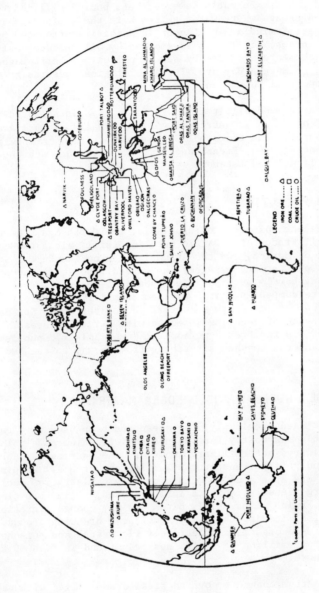

Fig. 32   Representative world ports capable of accommodating 150,000 DWT vessels.

FIGURE 33

ESTIMATED GROWTH IN TONNAGE OF THE
HIGH SEAS FISHING FLEETS OF SELECTED NATIONS

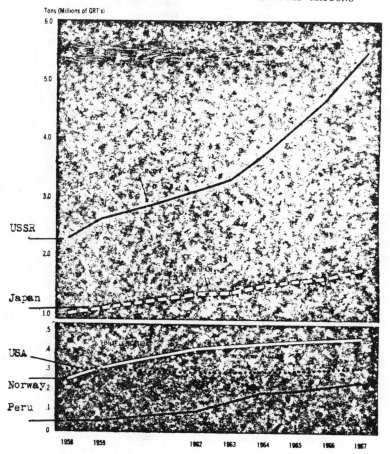

Ships larger than 5 net tons operating in the open seas and major inland
seas such as the Caspian and Black Seas and the Great Lakes

NOTE: Tonnage scale changes for Japan and Soviet Union.

Sources: Bureau of Commercial Fisheries, Bulletin of Fishery Statistics.
No. 14, Food and Agricultural Organization of the United Nations.
1966; Embassy of Peru; Miscellaneous.

FIGURE 34

MERCHANT FLEETS OF THE WORLD

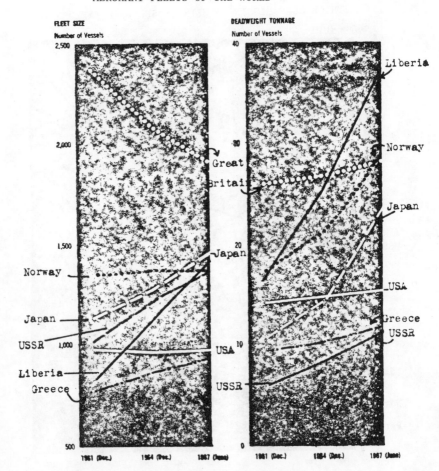

Oceangoing steam and motor ships of 1000 gross tons or more. Excludes
ships operating exclusively on inland waterways and special ships such
as channel ships, icebreakers, military ships, etc.

Only privately-owned U.S. ships are included. U.S. government ships,
excluding the reserve fleet, rose from 64 vessels in 1961 to 201 vessels
in 1967 but this largely reflects activation of reserve ships for war duty.

Sources: Merchant Fleets of the World, 1961, 1964, 1967, Maritime Administration, U.S. Department of Commerce

FIGURE 35

GROWTH OF LARGE MERCHANT SHIPS
OVER 200,000 DWT IN THE WORLD

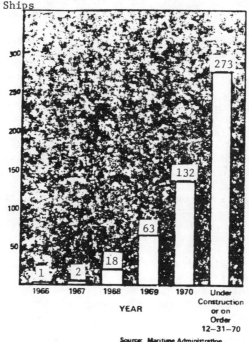

Ships

Source: Maritime Administration
         U.S. Department of Commerce

Yet, organic wastes placed in the upper water layers where
light penetrates, and properly diluted, could fertilize the area;
and warm discharge waters conducted to lower ocean layers could
buoy up nutrient rich water from the deep to the surface waters.
Careful dumping may occasionally be favorable or permissible.

A Delaware based non-profit organization, Environmental Concern, Inc., constructed from 1971 through 1975 along the U. S.
East Coast, salt marshes using fine particle, uncontaminated sediments from dredge spoil. Within three months of pumping and
seeding with indigenous growth, the marsh is stabilized though
no bulkheads have been built. Occasionally seeds will not germ-
inate because of an hypersaline habitat; using plant material
triples the cost. Heavily polluted sediments, however, can return
heavy metals to the water column.

Further studies are examining the use of dredged sediments as
substrates for oyster and clam beds. At the University of Rhode
Island it was suggested that polluted spoil be buried at sea and
to cover it with a 50 centimeter layer of unpolluted material.
Horizontal leaching is insignificant and vertical rise of pollu-
tants impossible. The technique requires high precision navigation for siting and dumping, and locations free of erosion and
interference with ocean uses.

Dredged material may be used to build artificial islands, as
was done in San Diego harbor where additional tie-up facilities
were needed. However, these may produce hydraulic dislocations
upsetting wave action and silting rates.

The international convention on ocean dumping agreed upon in
London in September 1972 gives a list of blacklisted products and
graylisted materials; the latter, may be dumped under certain controlled circumstances. However, signatory nations are still few.

4. REGENERATIVE POWER OF THE OCEAN

Some ten years ago, the ocean was praised not only for the
promise it held to solve some of humanity's problems, but also
because it provided an ideal dumping ground for both human wastes
and radioactive wastes. Since then, concrete containers containing nerve gas, ypresite, and others with radioactive wastes have
been thrown into the ocean; so have street-cars, cars, untreated
sewage and used water. The list is far from comprehensive. Yet,
in fine, air pollution, land pollution and water pollution are
all ocean pollution. The ocean is endangered ecologically, chem-
ically, physically. We cannot allow further deterioration of the
ocean. Furthermore, the ocean does not destroy matter as rapidly
as was once thought. The recent experiments conducted on food

retrieved from the submersible Alvin refloated after one year at
the bottom shows that conservation in the ocean is quite surpris-
ing. The sea has favored the biological balance by absorbing refuse and diluting the substances which could have been destructive.
Through the action of the marine flora and fauna, of absorption,
digestion, transformation and concentration phenomena, regenera-
tion of the vital environment favorable to life upon our plant
takes place, but the noxious elements cannot be allowed to exceed
the regenerative capacity of nature. Until this century, nature's
capacity has never been challenged, but now it is endangered. The
challenge has taken on new dimensions because, in addition to the
large quantities of wastes brought to the ocean by rivers or directly thrown in at the coast, and the chemical refuse and nuclear
wastes poured into the ocean, we are faced with petroleum wastes
and the cleansing of ships at sea. Soviet researchers found hydro-
carbon derivatives at depths of 100 meters in the Baltic Sea and
a nature reserve near the Hange peninsula would be certainly destroyed if a planned refinery were to be constructed in the area.
As far as some outspoken ecologists are concerned, either we stop
exploration and drilling for marine-derived hydrocarbons, or we
will founder and die in a polluted quagmire.

5. OIL POLLUTION

The petroleum industry has made serious efforts to combat
accidental spills, and to prevent them, in offshore drilling. Out
of some 20,000 wells drilled in American waters, there have been
only four serious accidents in the last quarter century, none of
which caused permanent damage to the environment. Improved technology, including electronic sensors, should help make that average
even better. The National Academy of Sciences estimates that offshore drilling and production account for only just over 1% of the
oil in the ocean.

In addition to industry controls, the U. S. government keeps
close watch on each platform's discharge waters. Regulations
permit an operator to discharge an average of no more than 50
parts of oil to a million parts of water from its producing operations. That's a little like diluting an ounce of Scotch with 150
gallons of soda. In either case, nobody gets very polluted!

But an ailing economy may lead to curtailing of environmental
safeguards. Russel E. Train, administrator of the United States
Environmental Agency, stated in New Orleans in April 1975 before
the Audubon Society, "Dredging and development is destroying the
marshes ... some 200,000 acres of shallow coastal bays in the
Gulf and south Atlantic areas [have been wiped out] over the past
two decades. Chemicals and sewage and oil spills are slowly and
steadily sapping the ocean's ability to serve as a well of life..

... the seas ... have become a sink for enormous quantities of chemicals from fertilizers, herbicides and pesticides used in agriculture far inland ... We have followed a policy of plunder-now-and-pay-later whose price tag must all too often be paid by victims far from the scene of the crime." These statements sound a prelude to strict improvements measures, yet one cannot but ponder the ominous omen in the words of President Gerald Ford in the Fifth Annual Report of the Council on Environmental Quality: "In my judgement the course we must continue to follow is a policy of trade-offs between economic and ecological realities. The environmental movement has matured enough to go along with compromises requiring to give up the freedom to decide what to do with one's property, and in order to get more energy to accept oil spills, strip mining and uncontrollable environmental problems."

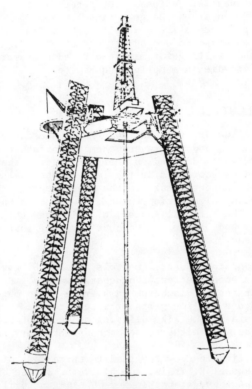

With elevating legs, the jack-up rig can be floated to location and then raised or jacked up on the legs to appropriate height above water. This rig is normally limited to about 300-foot water depths.

V

WHO OWNS THE OCEANS?

A DISTINCTION must be made between maritime law and the Law of
the Sea. The first is a body of laws dealing with commercial and
admiralty matters, while the second is an attempt to regulate
political aspects and economic problems pertaining to the ocean
as national and international territory. The Law of the Sea has
evolved from both custom and treaty. It is an area of international law which attempts to provide order in the marine environment and which has been practically universally accepted for over
three centuries.

1. THE FREEDOM OF THE SEA

The Dutch philosopher Hugo Grotius declared that the sea is
open to use by all because no part of the sea may be regarded as
belonging to any given nation. The idea proponed in 1607 was
sound for the times. This "rule" was tempered gradually by indicating that the actual territory of a State extended three miles
beyond the water line, three miles being by the 1800s the firing
range of coastal batteries; yet, even within that zone, all ships
retained the right of innocent passage.

The first challenge to the three-mile limit came from czarist
Russia, at the beginning of the 20th century, when it unilaterally claimed a 12-mile exclusive fisheries zone, and the Soviet
successors declared that this 12 mile zone was in fact a territorial sea. True, it was not until recently that the international
accord became an international discord as a result of a new
assessment of the ocean's value, advance in technology, population
growth and political changes. The freedom of the sea has been
threatened ever since.

In 1947 President Truman affirmed a claim to all resources
of the continental shelf of the United States, thus extending in
fact the national territory along the sea bottom but not including the water column above it. Chile, Ecuador and Peru used this
measure, in 1952, to claim sovereignty and jurisdiction to an
area extending 200 nautical miles from their coast; they have
hardly any continental shelf, but their waters are rich fishing

grounds. The United States established its 200-mile limit in
1977.

We have shown that the surface of the sea, the water column
and the sea bed are very valuable. They are traffic lanes, produce energy, contain life and mineral resources. Ocean exploitation led to the first International Law on the Uses of the Ocean
meeting in Geneva in 1958. None of the four conventions actually
settled the matter of the breadth of the territorial sea. Another
attempt in 1960 ended in failure.

2. INTERNATIONAL REGIME PROPOSALS

Several proposals dealing with the ownership of the ocean, beyond territorial waters, as well as with the extent of territorial
waters, have been placed before the United Nations. Private firms
are hesitant to invest large sums required by prospecting and ex-
ploiting of ocean areas as long as ocean ownership is not clearly
defined. They can hardly be blamed since several have suffered
economic losses by sudden nationalizations which, in several instances, were outright confiscation. Some companies have announced
definite plans for the coming years to start mining in the Pacific
Ocean as deep as 5,000 and 6,000 meters.

Japanese commercial enterprises are backed by the government
and an increase of 15% was decided in 1972 over the preceding
year's budget for mineral exploitation of the ocean. But Japan
is by no means alone in its expanding interest in the ocean, the
efforts of the Soviets have been widely publicized and the United
States has been in the forefront of oceanographic activities for
some time. It was inevitable that legal problems would arise from
positions which are unavoidably divergent. There is virtually no
marine law jurisprudence; there is no clear definition of the
boundary separating two states at sea. The only international
agreement with some relevance to present conditions is more than
ten years old and, since then, technological progress has outdis-
tanced the legalities. At that time, the continental plateau was
defined as the bottom of the sea and the sub-bottom layers of the
adjacent marine regions, located outside the national territory,
up to a depth of 200 meters, or beyond that limit if "the depth of
the overlying waters permitted the exploitation of the natural
resources of said regions." This satisfied the signatories of the
convention because no techniques existed then which would permit
penetration at greater depths. It left virtually without "marine"
territory countries bordering areas where the continental plateau
is non-existent. An example is Peru which is asking, in compensa-
tion, territorial limits of 300 km, and patrols its "territory"
in consequence, often capturing foreign vessels and demanding
heavy fines. U. S. fishing vessels often fall victim of this
practice and the U. S. Government, which refuses to recognize the

Peruvian limits, bails them out. However, some countries, which
have neither the manpower nor the financial means of enforcing
respect of their territorial limits, could quite well entrust this
task to third parties and thereby upset political situations.
Finally, must we ignore the objections of landlocked countries
which, perhaps rightfully, feel that the ocean belongs to all humanity and demand their share of the wealth of the "last frontier
on earth"?

Discussion about international law for the economic uses of
the sea started when the U. N. delegation from Malta suggested
that the ocean should not be up for grabs but should instead be
shared by all of mankind. Vociferous objections were instantly
raised against all proposals in this direction, especially from
the firms which have been-tapping the mineral resources of the
sea, oil and gas particularly.

A properly established international regime, Ambassador Pardo
pointed out, "contains all the necessary elements which should
make it acceptable to all of us here: rich and poor countries,
strong and weak, coastal and landlocked States".

In conclusion he proposed, as "our long-term objective", the
creation of a special agency with adequate powers to administer in
the interest of mankind the oceans and the ocean floor. "The agency should be endowed with wide power to regulate, supervise and
control all activities on or under the oceans and the ocean floor."
It should be based on the following principles:

"1. The sea-bed and the ocean floor, underlying the seas beyond the limits of national jurisdiction as defined in the treaty,
are not subject to national appropriation in any manner whatsoever."

"2. The sea-bed and the ocean floor beyond the limits of national jurisdiction shall be reserved exclusively for peaceful purposes."

"3. Scientific research with regard to the deep seas and
ocean floor not directly connected with defence, shall be freely
permissible and its results available to all."

"4. The resources of the sea-bed and ocean floor, beyond the
limits of national jurisdiction, shall be exploited primarily for
the needs of poor countries."

"5. The exploration and exploitation of the sea-bed and
ocean floor beyond the limits of national jurisdiction shall be
conducted in a manner consistent with the principles and purposes
of the United Nations Charter and in a manner not causing unneces-
sary obstruction of the high seas or serious impairment of the

marine environment."

Further fundamental principles were suggested:

"1. The sea-bed and the ocean floor are a common heritage of
mankind and should be used and exploited for peaceful purposes and
for the exclusive benefit of mankind as a whole. The needs of
poor countries, representing that part of mankind which is most in
need of assistance, should receive perferential consideration in
the event of financial benefits being derived from the exploitation of the sea-bed and ocean floor for commercial purposes."

"2. Claims to sovereignty over the sea-bed and ocean floor
beyond present national jurisdiction, as presently claimed, should
be frozen until a clear definition of the continental shelf is
formulated."

"3. A widely representative but not too numerous body should
be established in the first place to consider the security, eco--
nomic and other implications of the establishment of an international regime over the deep seas and ocean floor beyond the limits
of present national jurisdiction; in the second place, to draft a
comprehensive treaty to safeguard the international character of
the sea-bed and ocean floor beyond present national jurisdiction;
and in the third place to provide for the establishment of an
international agency which will ensure that national activities
undertaken in the deep seas and on the ocean floor will conform
to the principles and provisions incorporated in the proposed
treaty."

In December, 1968, a U. N. permanent Committee composed of
42 states, was instructed (1) to study the legal principles and
norms that would promote international cooperation in the exploration and use of the sea-bed and the subsoil beyond the limits of
national jurisdiction; (2) to study the means of encouraging the
exploitation and use of the resources of this area in the light
of foreseeable technological development and economic implications,
bearing in mind the fact that such exploitation should benefit
mankind as a whole; (3) to review and stimulate the exchange and
widest possible dissemination of scientific knowledge on the subject; (4) to examine proposals to prevent marine pollution that
may result from resource exploration and exploitation.

The Twenty-fifth General Assembly of the United Nations
adopted a Declaration of Principles which embodied the proposals
of 1967 and elevated the principle of the common heritage of mankind to the status of international law.

3. THE UNITED NATIONS AND THE "LAW OF THE SEA"

Today it is obvious that legislation regarding the sea is no longer adequate. When the agreements were reached in Geneva in 1958 no one realized that nations would ever want to exploit the sea at a greater depth than 200-300 meters. But today it seems quite possible that by the year 1980 the oceans could be exploited to depths of 4,000 meters. However, legislation means restriction. The idea of an international sea-bed regime means limiting national jurisdiction and the question is where to set the limits and over what.

3.1 THE PROBLEMS TO BE SOLVED

In this respect there are several problems to be solved; for instance, the problem of the continental shelf, which has not yet been defined; territorial sea and zone, fishing banks, innocent passage, international regulation and regime, exploitation and conservation of resources, and measures against pollution. Furthermore, since World War II, the fishing fleets of the world have been renewed and changed radically and catches increased rapidly. There is reason to believe that fishing might start to decline in the near future. Oil and gas now represent 90% of all the mineral wealth extracted from the sea, but today we see signs of exploitation of other minerals as well.

There is widespread belief that an international regime is necessary in order to avoid clashes over the riches of the sea but the world's nations have quite different views of the form of such a regime. Some want to give the authority in this matter to the United Nations or to the International Court in The Hague, but experience has taught that these institutions have no means of enforcing their decisions and thus must have the consent of all parties involved in order to achieve agreement.

There seems to be common agreement that an international regime of the sea must encourage peaceful economic exploitation of the sea and that it must ensure maximum benefits, not only for coastal states, but for the entire international community. Such a regime must furthermore be flexible enough to be adapted easily to new technology. The Geneva agreement in 1958 was very flexible at that time, but signatories did not expect that it would shortly be possible to exploit resources at great depths.

Committees, preparing the conference of the United Nations on the Law of the Sea (1973), discussed topics such as freedom of passage through straits, coastal states' preferences as regards fisheries, international regime versus national jurisdiction, conservation, and scientific research.

3.2 PROPOSALS FOR INTERNATIONAL MANAGEMENT

Various types of proposals concerning international machinery were put forward. The four main types are:

- International machinery for information exchange and preparation of studies.

- International machinery with intermediate powers.

- International machinery for licensing and registration.

- International machinery with comprehensive powers.

Many nations made proposals: the United States, Malta, Tanzania, Latin American countries, France, Britain, the Soviet Union, Canada, a group of landlocked and shelf-locked nations, and others. Naturally, numerous theses are put forward. Among these, one position insists on the rights of first occupancy, which would grant ownership to the first discoverer, going as far back as the 15th century when new lands were annexed by European nations. This would unavoidably lead to armed conflict and guarantee possession to the most powerful nations. The world ocean would soon be divided among these and become a private lake. Another thesis is based upon the logical continuity of the coast and the continental slope; hence, the national territory would continue underneath the sea, as long as the sea bottom would be without fracture or break the extension of the coast. Still another position recommends the subdivision of the entire ocean bottom, beyond the continental plateau, among all countries. One point that remained unsettled in this last plan was whether all shares would be equal, or which rules would determine the various shares. A final proposal would apply the high seas rule to the ocean bottom, and thus all countries would own, jointly, the ocean bottom while no nation would acquire an individual property title over a given area.

With joint ownership, new problems arise. Exploitation would then be done either by the community of nations or else the community would grant concessions for which research permits and exploitation permits would be granted in exchange for the payment of license fees. Were the community itself to attempt exploitation, the ultimate outcome would lead to a colonization of the oceans by the superpowers, since they alone possess a sufficiently advanced technology. On the other hand, a system of licensing with payment of fees, would theoretically benefit the entire community, with each country getting a share of the fees even though the work would be done by highly developed countries.

The American proposal would substitute for a regime of

laissez faire, of which the United States was at one time the
determined supporter, a supranational collaboration which would
not hamper freedom of navigation on the high seas and beyond
reasonable territorial limits. This proposal recognizes the overall interests of all humanity, the need to protect the environment,
and the obligation to fairly apportion the benefits reaped from
the ocean. This plan can be summarized as follows: first, beyond
200 meters depth, resources extracted from the ocean would be used
partly for scientific research and partly for improvement of the
economic conditions in the underdeveloped countries. Respect of
the convention and administration of a common fund would be en-
sured by an international authority, which would have the necessary instrumentalities to impose rules enforcement, e.g., control
services, courts, parliamentary assemblies, political councils and
technical commissions.

The administrative domain of this authority would encompass
all supra- and extraterritorial waters when they have been defined.
This international authority would also grant exploration and exploitation licenses, pass anti-pollution regulations, and, in
general, be the watchdog over the aquatic environment.

Although national sovereignty would affect areas of less than
200 meters depth, national sovereignty would nevertheless be somewhat reduced. The coastal state would have exploitation rights in
its national area but, nevertheless, would have to pay the international authorities some fee, based on the product harvested.
This fee could be as high as two-thirds of the sum collected at
the time the license was granted. The proposal provides special
consideration for islands and coastal states very dependent on
fisheries.

This clause places riparian states under control of the international body and makes it impossible for them to extend their
territory into the international area. The greed and indifference
of coastal states would thus be constrained. Such nations would
be prohibited from transforming their "maritime territory" into
pollution factories. The main opposition to the American proposal
is the Tanzanian proposal, which is favored by many developing
countries. Tanzania wants each state to set its own limitation
of territorial waters with resources outside the national jurisdiction distributed according to a rule which states roughly "to
each country according to its needs". Members of the U. N. shall
provide money on a basis proportional to contributions to the
United Nations, which means that the United States and Europe
would bear the burden. Then, income would be used for administration, exploration and exploitation expenses. The remainder, if
any, would be distributed to the states that belong to the U. N.
in inverse proportion to what they contribute to support the U. N.
Thus, when there was an income, it would be used at first to pay

back the sums paid to this international body and the remainder
shared by all, with the poorest nations getting the largest share.

The Soviet proposal favors creation of an international authority to administer the sea bed resources. This authority would
be governed by a small executive committee and controlled by an
assembly of the signatory states.

This executive committee would be empowered to enforce adher-
ence to the terms of the treaty, control industrial activities,
assess reserves, distribution, and localization of resources, deliver licenses, and determine disposition of benefits. On the
other hand, this international authority would not have jurisdiction over the sea bed nor the immediately underlying layers, nor
would it have exploration and exploitation rights.

These two restrictive clauses do not appear in the Tanzanian
proposal, which would authorize the international authority to
own mining equipment, establish research and oceanographic institutes, and provide experts and technological assistance to developing countries which would become active in ocean exploration and
exploitation. The international authority would be empowered to
set prices and to decide the quantity of products that may be
offered for sale, thereby protecting the smaller nation's economy.

The Latin American group has come up with a proposal similar
to Tanzania's. It does not foresee an international machinery nor
the right for such an authority to grant licenses; nor does it
vest in anyone de facto or de jure ownership of the sea bed. The
general trend is to put territorial limits at 20km, with the right
of innocent passage for any vessel, and economic jurisdiction at
300 km.

A proposal which has gathered strong support calls for the
creation of international machinery with comprehensive powers.
This receives the widest consideration from the U. N. General
Secretariat, and appears in proposals from the United States,
France, and Great Britain. Yet the two most powerful countries -
the United States and the U.S.S.R. - seemingly agree to support
an extensive international regime for the oceans, with modest
claims of national jurisdiction. The American proposal has received much criticism from large international consortiums which
claim that it would slow scientific research and exploration, and
would foster exploitation conflicts. While this argument is not
without merit, private concerns managed to find, and exploit,
mineral resources on land, notwithstanding legislation that was
quite restrictive; as a matter of fact, exploitation was so an-
archic that reserves are often exhausted and the countryside is
left with scars; the land is polluted. Perhaps a slower paced
exploitation might be wholesome and benefit future generations.

Unfortunately, progress is very slow and probably no treaty will be signed before 1980. The workload is heavy since so many international conventions must be re-examined. Some examples are: territorial sea, contiguous zones, high seas, continental platform, fisheries, and conservation of high seas' biological resources. It is difficult to be optimistic when one realizes that of the 149 states that are members of the United Nations organization, only 28 states have ratified the 1958 Geneva Convention.

3.3 THE LAW OF THE SEA AND ENVIRONMENTAL PROTECTION

As mentioned under a preceding heading, the Law of the Sea is equally urgently needed to insure unwarranted pollution of the ocean. Developing countries view environmental restrictions as an encroachment upon their newly acquired sovereignty and interpret it as a veiled attempt by developed nations to hamper industrialization, witness the speech by Zaire's Mobutu in Washington: "if pollution means industrialization, I want pollution".

Although sovereignty gives the coastal State power to control marine pollution originating in areas under its jurisdiction, it cannot protect that State from pollution originating in areas under the sovereignty or jurisdiction of another State. Sovereignty may ensure that seabed hydrocarbons are exploited rationally in areas under coastal State jurisdiction but does not ensure harmonization of resource exploitation with other uses of the sea such as navigation. Furthermore, vastly extended coastal State maritime sovereignty inevitably hampers scientific research which is an essential prerequisite to resource management and conservation; it is also a threat to the international transport of hydrocarbons since any coastal State would have the authority to subject to arbitrary tolls or to crippling conditions the transit of energy carriers through straits and maritime areas under its sovereignty.

The International Maritime Consultative Organization (IMCO) and other UN agencies have provided, and continue to provide a framework for the negotiation of agreements designed to reduce environmental dangers resulting from activities, particularly navitation, in the ocean space but such agreements can only be developed slowly. They have comparatively few signatories and lack an effective enforcement mechanism. Furthermore, problems such as management of living resources, harmonization of ocean uses beyond national jurisdiction, reasonable freedom of scientific research or reservation of safe and secure transit for peaceful purposes cannot be effectively dealt with within the traditional parameters of international law.

Wherever coastal State sovereignty may, by the forthcoming U. N. Conference of the Law of the Sea, be held to end, it is

clear that some aspects of sovereignty in the marine environment
must be constrained and subject to international control. It is
equally clear that complete freedom in whatever marine area may
remain beyond national jurisdiction is too dangerous a policy.

3.4 PASSAGE THROUGH INTERNATIONAL STRAITS

One aspect of the potential new "Law of the Sea" that will
have great impact is the survival of the right of innocent passage.
Several countries may contest it as a result of the extension sea-
ward of national territorial limits which would place several
straits entirely within their jurisdiction. They include Gibral-
tar, Bosporus, Skagerrak and Kattegat, Magellan, Malacca. Retrout-
ing will increase costs of goods transported, and particularly
of oil.

4. AND NOW TOMORROW

Since the basic concepts of international law are no longer
valid, the international community must accept a new principle as
a basis for a new law of the sea. The only principle suggested so
far in international negotiations is that the seabed beyond national jurisdiction is a common heritage of mankind.

However, with the steady trend to roll States' boundaries
further out at sea, this common heritage is rapidly losing any
real significance. One may fear what exclusive sovereignty may
lead to.

Each use of the ocean must be at most a bailment, limited in
the period of use, regulated in management, requiring a standard
of care, measured in consumptive use and compensated by rent or
royalty.

Acceptance of the common heritage principle implies the creation of international institutions for ocean space as a whole.
This does not mean, of course, international administration if all
ocean space - a development which seems beyond the bounds of contemporary political reality - but rather international management
of ocean space and its resources beyond national jurisdiction and
effective international standards with regard to major uses of the
sea, whether within or outside national jurisdiction. Although
the international institutional framework must be global and comprehensive, it should probably largely work through regional mul-
tipurpose bodies. Finally, the new institutions would not be
complete without an impartial and flexible system for the compul-
sory settlement of disputes. Contemporary international law,
based on the twin principles of sovereignty and freedom cannot

adequately deal with the dangers involved. There must exist
agreed rules for harmonization of major uses of ocean spaces;
navigation itself cannot be entirely free in heavily trafficked
areas.

The Caracas session of the Third United Nations Conference on
the Law of the Sea has made substantial progress toward the establishment of international machinery, including an enterprise,
to regulate, manage, and develop one particular use of ocean space
beyond national jurisdiction and its resources, that is, the exploitation of resources from the deep floor of international ocean
space.

Little consideration, however, has yet been given to the
development of new forms of international cooperation with regard
to other ocean resources and to the traditional uses of ocean
space, such as navigation, which are being transformed by technological advance. Also, many new uses of the ocean are arising,
and interaction between both new and traditional uses of the sea
is growing. Increasingly, the need for the preservation of the
marine environment requires coordinated action. International
criteria for the harmonization of certain ocean uses must be ela-
borated and the uses themselves must be subjected to regulation
if conflicts are to be avoided, and if the potential benefits
which the oceans offer are to be realized.

A necessary complement to the exercise of comprehensive powers by coastal states in wide areas is international management
of ocean space beyond national jurisdiction (international ocean
space). The continued existence of the freedoms of the High Seas
beyond national maritime areas must frustrate to a greater or lesser degree national management of the sea and its resources within
national jurisdiction. This is particularly the case in the light
of contemporary military and technological developments. In addition, such international management would enable landlocked and
other geographically disadvantaged States to participate on an
equal footing in the management of benefits from international
ocean space and its resources.

It is in the long term interest of all States to seek to
realize the goal set at Oaxaca by the President of Mexico. It is
in the immediate interest of all developing nations striving for
a new international economic order to establish a new regime for
ocean space. The creation of a machinery to administer ocean
space beyond national jurisdiction and its resources would be an
important institutional step in the direction of the new economic
order.

A new strategy is suggested which, building upon the results
of Caracas, could provide a common focus for the work of the three

main conference committees. It could cement the unity of developing countries through adoption of a common goal, and could appropriately utilize international institutions.

The new stategy would be based upon the assumption that international management of ocean space beyond national jurisdiction
is a necessary complement to the exercise of comprehensive powers
by coastal states in wide areas. Accordingly, the establishment
of an International Seabed Authority as envisaged by the First
Committee of UNCLOS will need to be supplemented by other organizational mechanisms dealing with the managment and regulation of
other uses of international ocean space. At the same time it is
suggested that appropriate arrangements be made to deal with all
ocean space activities not covered by existing intergovernmental
organizations, particularly with regard to ocean space beyond
national jurisdiction.

Such arrangements would entail functional coordination and
possible restructuring of these organizational mechanisms. There
is a need for a permanent body to keep under constant review the
existing activities of the United Nations system relating to the
seas and oceans and to provide a forum for the discussion of
emerging problems relating to ocean space. In addition, other
functions of this permanent body could be:

1. To integrate the work of the agencies and organizations whose
    primary activities are directed towards the oceans;

2. To deal with all ocean activities not covered by existing
    intergovernmental organizations;

3. To harmonize interactions of multiple ocean space uses;

4. To promote cooperation between national and international
    management systems;

5. to ensure effective international cooperation with technolo-
    gically less advanced countries in the development of national
    ocean space;

6. To ensure equitable distribution of benefits derived from the
    exploitation of the resources of international ocean space;

7. To promote the progressive development of the law of the sea;

8. To assume some functions with regard to dispute settlement.

5. UNRESOLVED QUESTIONS

The Exclusive Economic Zone (E.E.Z.) under consideration would grant a State very broad rights. It would be sovereign in matters of resources exploitation, pollution control, scientific research, and some thought is even being given to ship safety standards and operating personnel qualifications. If the concept appears to have been accepted, the "content" of the E.E.Z. has not been settled. Does it extend to 200 miles, or further? Does it include the water column or is it limited to the geological seabed? To Latin American countries there is identity between territorial waters and the E.E.Z., while other views differentiate between a territorial sea and the adjacent E.E.Z. in which, for some, limited sovereign rights can be exercised, and for others, not sovereign but preferential rights exist. Then, there is a group of countries that feels that, on the sea floor, they still possess the rights that were recognized in previous conventions, for the continental shelf. These continental shelf rights would be theirs in addition to those they would enjoy under the new Law of the Sea treaty for the E.E.Z. Finally, each nation, within its own waters and other States's waters, promises to combat pollution. But who is to police?

It is not improper to wonder if the new legislation for ocean space will be actually effective. While a State, in its sovereign water area is expected to keep a vigil on such matters as safety and pollution abatement, nothing is said about the State's ability for damages caused to areas outside its national jurisdiction, whether in international waters or another State's waters. We are all only too keenly aware that pollution of various types caused by ships pose a most difficult, and often most delicate problem. The London convention on ocean pollution, and its amendments, would be strictly enforced. Yet, years later, some signers of these agreements don't enforce its terms and scores of States have not even bothered to sign them. Will the Law of the Sea now under consideration fare any better?

The United Nations Environmental Protection Agency (UNEP) does not impose policies; it merely provides a forum for sovereign States to enter into discussions. Nor does UNEP make any claims of monopoly on environmental efforts. It differs, however, from other agencies involved in environmental matters in that their primary interest is not exclusively environmental. Since UNEP, though only modestly staffed, has funds, disposes of the necessary mechanism, and can organize conferences, it might perhaps be the medium through which effective antipollution covenants may be agreed upon and inserted in the provisions of the forthcoming Law of the Sea.

Will negotiations to restrict the military uses of the ocean

lead to agreements to be embodied either in the Sea-Bed Treaty
through amendments, or in special protocols to that Treaty or to
the new Law of the Sea? For Alva Myrdal, points to be negotiated
should include extension of the Sea-Bed Treaty to cover all national military installations, including platforms, on, or moored
to, the seabed, prohibition of the testing of missiles, and all
maneuvers over the high seas, agreements to account publicly for
mobile detection systems, and to prohibit dumping of nuclear and
chemical weapon wastes. Such international agreements should be
completed by regional agreements as to the military use of certain
oceanic regions, including the questions of limiting rights of
passage, bunking and base establishment, for certain types of
vessels, particularly those carrying nuclear and chemical weapons.

Will the creation of an Exclusive Economic Zone be a very
beneficial solution for mankind? Will developing nations be able
to exploit their E.E.Z. and to safeguard it against pollution?
Some believe that the creation of such E.E.Z. would benefit mainly the developing countries, which he maintained would be to the
advantage of mankind, while presently, for instance, fifteen countries get 75% of the world's fish catch. To the position that
the Freedom of the Seas issue is used to exert pressure on weaker
countries, some counter that no thought was given in Caracas to
the real silent majority - the unborn generations - and instead
an incredible thirst was displayed for the exercise of sovereignty
predominently on the part of energy nations.

If the desire to have international ocean space viewed as
a unity is close to unanimous, territorial claims and sovereignty
rights are all but unified. Nor do we seem to have come much
closer to an agreement on ocean resources exploitation.

Obviously if the economic zone proposal would come to pass
into law, the celebrated "common heritage of mankind", originally
seen as taking in 75% of ocean space, would be shrunk to cover
merely depth zones where, for the next decades at least, resources
are pretty well limited to polymetallic nodules. With each individual State responsible for pollution abatement and ocean exploitation in its own broad economic zone, common heritage exploitation would be meaningless in the face of an array of frag-
mented organizations and competences vying for authority in international ocean space.

The Law of the Sea should ensure the maintenance in perpetu-
ity of the ocean living resources worth at present nearly $12,000
million annually, many of which are already over exploited.

Intensive fishing and pollution lend urgency to the need for
international arrangements for prompt and protective action. Presently exploited resources including whales, tunas and salmons,

occur within the proposed 200-mile exclusive economic zone and
move between zones. The new Law of the Sea will need to take account of this mobility while multilateral agreements, will have
to keep in mind that the potential yield from these resources is
limited to, at most, a doubling of present catches.

There is need for pollution control and monitoring, functional cooperation between national and transnational fisheries management, multinational cooperation in scientific research, and possible international licensing for international navigation. The
radical change in ship size poses new and complex problems including smaller crews but unchanged social hierarchy, inadequate
training facilities for automated systems use, a transformation
in traffic patterns, creation of new ports and their impact upon
communities ashore. All these problems require an extra-national
approach if they are to be solved and if accidents and stresses
are to be minimized.

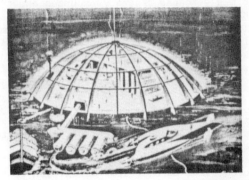

MARION SENYK

VI

CLOSING REMARKS

SOME of the thoughts expressed by Elisabeth Mann Borgese in "The Drama of the Oceans" (New York, Abrams, 1976) might appropriately be quoted here. The ocean as the source of life and our last terrestrial frontier is gravely endangered, principally because each man, each industry, each nation "takes out of and puts into the oceans what profits·him most today or at the latest tomorrow"; irrational and careless "marine exploration and exploitation may disrupt delicate systems of communication and cycles of reproduction, killing off many species ... Pollution and radioactive wastes may eventually destroy the phytoplankton in the upper layers of the Sea, which produces more than half the world's oxygen", with protentially catastrophic consequences. As many other writers and teachers, she urges the establishment of an international institution to manage ocean spaces, resources and their interactions, an integrative use of the oceans so uses enhance one another, minimizing wastes to abate pollution, development of aquaculture, aquaforming and conservation policies, monitoring of discharges and their limitations, and of a jurisdiction for surface and underwater vehicles and for the operations of oilmen, energy engineers and miners to ensure safety and prevent pollution and depletion.

The attention of political scientists, economists, geographers, biologists, engineers and other scientists is thus to be directed to four complexes of questions.

1. The need for a more precise definition of the scope of national jurisdiction. Its limits, in the oceans, depend on a precise definition of the territorial sea and the exclusive economic zone, on precise rules with regard to the drawing of straight baselines, and on either the absorption of the continental shelf concept in the exclusive economic zone or on its precise definition on the basis of clear and unambiguous geophysical characteristics of the seabed to be determined through an impartial judicial mechanism. Marine jurisdiction of Island States and of islets, reefs sandbanks, must be clarified, or else the incredibly reduced international area of the seabed will lose significance.

2.   Problems of effective management of national ocean space
in its relations with international ocean space and vice versa.
National ocean space comprises both the Territorial Sea and the
Exclusive Economic Zone wherein the Coastal-States have comprehensive powers of resource management. These powers must be
matched by equally comprehensive powers vested in an international
organization with regard to that area of ocean space which remains
beyond national jurisdiction. International areas would, by bene-
fiting mankind as a whole, reduce inequalities between nations
while the international institutions could promote simultaneously
scientific research and transfer and technological know-how.

3.   The need to take the scientific research and the implications of technological advance more fully into account in the ela-
boration of the Law of the Sea. Increasingly essential to the
needs of multiplying populations, the rational development of
ocean resources is dependent upon scientific research which must
be subject to sane regulation and promoted throughout ocean space.

4.   An ocean regime must encompass the oceans as a whole and
a sub-system of a global system. Jurisdictional decisions, including those affecting the Exclusive Economic Zone, must reflect
that paramount concern. It is not a matter of geographical re-
alignment or of partition; the marine ecosystem does not correspond to any demarcations. Nor is mankind for which the "common
heritage" is prescribed confined to coastal state nor indeed to
the present generation.

The rapid scientific and technological developments which
revolutionized the conventional uses of the sea call for management as the only alternative to conflict and possible catastrophes.
The nature and possible activities on the seabed cannot be dis-
sociated from the water column, the surface and the atmospheric
interface and must be considered part of any management system;
claims for national jurisdiction carry a surrogate responsiblity
in that management. Any Law of the Sea which does not respect
and embody those overriding considerations will prove to be in-
effective if not inoperable.

Contemporary technology has improved to such an extent that
operations have undergone a change in scale; the number of activities has substantially increased and so has the extent of the
ocean areas subjected to man's action. The traditional Law of the
Sea was designed for a light and quite limited use of the Sea;
however, the expected exploitation of the entire globe will lead
to a new civilization which might, no longer, be landbased.

Yet global problems are neglected. "The walls of Jericho
fell at the blast of a trumpet, but no trumpeting ever built a

wall. In other words, all criticism is fine, but the task at hand
is to decide what the world is going to do about the unfolding new
situation."

Submarine builder John H. Perry, Jr. stated some time ago,
"Our oceans are polluted. Fish life is contaminated and depleted.
Recreational beachfronts are disappearing. Hurricanes spawn at
sea and rage uncontrolled over the land. The Gulf Stream, the
world's greatest 'river' is an energy source untapped at a time
when an energy crises is upon us. Private capital cannot do it
all". And Perry, who was a member of the late U. S. President
Johnson's commission to prepare a national ocean program which
led to the creation of NOAA, cautioned, "Development of resources
in and under the oceans could help overcome the acute problems of
domestic inflation and imbalance of foreign trade payments. Every
ton of minerals and seafood and every barrel of oil we import ag-
gravates these problems. The nation's imports could be reduced
drastically with full development of food and other marine resources".

The exploitation of the ocean for the benefit of all humanity
must be conducted as part of an overall plan that provides the
clean-up of the marine environment, the fight against further pollution, the improvement of existing conditions, and the protection
of the environment. Whether we consider chemical, biological, or
geological resources, ocean exploitation should not be allowed to
take place amidst juridical dispute, moral irresponsiblity or at
the expense of future generations. Richard Cowen once compared
the man who lives from what the earth provides as living off his
capital, while the man who lives from the exploitation of the
oceans is like a man living off his dividends. The capital-ocean
must be well cared for.

Bibliography

Alexander, L. M. - 1968
    Geography and the Law of the Sea: Annals, Assoc. of Amer.
    Geogr. 58, 1, 177-197.

Anderson, J. H. and Anderson, J. H., Jr. - 1965
    Power from the sun by way of the sea?: Power, Feb. 63-66.

Anfhammet, I. W. and Deichemann, W. B. - 1971
    Abstracts and summaries of the literature on drugs from the
    sea, 1967-1970, University of Miami Technical Bulletin
    (Sea Grant Institutional Program) no 16, May.

Anonymous - 1967
    President's science advisory committee on the world food
    problem: Commercial Fisheries Review, Aug.-Sept., 1-3.

Anonymous - 1968
    Marine Science Affairs: Washington, U.S. Govt. Printg. Office.

Anonymous - 1969
    Our nation and the sea: Washington, U.S. Govt. Printg. Office.

Anonymous - 1971
    Ocean mining comes of age: Oceanology International, VI, II,
    34-41, VI, 12, 34-38.

Anonymous - 1969
    Puerto Rico - Water Resources Research Inst., Parsons Company,
    Offshore airport planning: Los Angeles, Ralph M. Parsons Co.

Anonymous - 1970
    Economic aspects of ocean activities; Economic factors in the
    development of a coastal zone: Cambridge, Mass. Inst. of
    Technology.

Anonymous - 1974
    Who rules the waves?: Energy Policy, (Sept.), 268-269.

Anonymous - Coastal Plains Center for Marine Development; Develop-
    ment activities in the marine environment of the coastal
    plains region: Washington, Coastal Plains Regional Commission.

Anonymous - 1974
    Inter-University Program of Research on Ferromanganese Depo-
    sits of the Ocean Floor. Phase 1 report. April 1973, Wash-
    ington, D.C. Sponsored by Seabed Assessment Program, Inter-
    national Decade of Ocean Exploitation. National Science
    Foundation.

Arehart, J. L. - 1969
Oceanic drug chest: Sea Frontiers, 15, 2, 99-107.

Aubert, M. and Aubert, J. - 1973
Pollution marine et aménagement des rivages: Nice,
C.E.R.B.O.M.

Aubert, M. - 1971
Cultiver l'océan: Paris, Presses Universitaires de France.

Auburn, F. M. - 1972
The 1973 conference on the Law of the Sea in the light of
current trend in State seabed practice: Canadian Bar Review
L, 87-109

Bardach, J. - 1968
Harvest of the sea: New York, Harper & Row

Beau, Ch. and Nizery, M. - 1952
Utilisation industrielle des differences de températures
entre les eaux profondes de la mer et les eaux de surface:
Fourth World Power Conference, Proceedings, IV, 1-12.

Berryhill, H. L. - 1974
The worldwide search for petroleum offshore. A status report (1947-72): U.S. Geolog. Survey Circ. 694.

Bischoff, J. L. - 1969
Red Sea geothermal deposits. Their mineralogy, chemistry and
genesis: in: Degens & Ross, ed., Hot brines and recent
heavy metal deposits in the Red Sea: New York; Springer,
368-378.

Borgese, E. M. - 1976
The drama of the oceans: New York, Abrams.

Boulding, K. E. - 1964
The meaning of the Twentieth Century: New York, Harper &
Row, 141-143.

Carter, L. J. - 1968
Deep sea bed: who should control it?: Science 159, Jan. 5,
66-68.

Charlier, R. H. - 1970
Harnessing the energies of the ocean: A postcript. Marine
Technology Journal, IV, 2, 63-65.

Charlier, R. H. - 1970
    La faim des Kilowatts: Revue de l'Université de Bruxelles,
    Juillet, 1-17.

Charlier, R. H. - 1970
    Pollution problems. Give the earth a chance: Int. J. En-
    vironm. Studies, I, 129-139.

Charlier, R. H. - 1971
    The transatlantic telegraph cable and the "Physical Geography
    of the Sea": Act. Int. Congr. Hist. Sci. 12, VII, 73-76.

Charlier, R. H. and Vigneaux, M. - 1973
    France littorale: Apprehension géologique et stratégie
    d'aménagement: Bull. Inst. Geol. Bas. Acquit. 14, 181-192.

Charlier, R. H. and Vigneaux, M. - 1974
    Towards a rational use of the oceans: Proc. U.S. Naval In-
    stitute, 100, 4, 854, 26-41.

Charlier, R. H. - 1975
    Power from the tides. Comment: Nav. Eng. Jl., 87, 3, 58-59.

Charlier, R. H. and Vigneaux, M.- 1975/76
    The budding environmental clean-up (A viewpoint): Int. J.
    Enviromn. Studies, VIII, 39-52, IX, 39-52.

Christy, F. T., Jr. and Scott, A. - 1965
    The common wealth in ocean fisheries: Baltimore, John Hop-
    kins Press.

Christy, F. T., Jr. - 1968
    Protein concentrate and a rationalized fishery: Proc. Conf.
    on Alternat. for Balanc. Fut. World Food Prod. and Needs.

Clark, J. - 1974
    Coastal Ecosystems: Ecological considerations for manage-
    ment of the coastal zone: Washington, Conservation Foundation.

Clark, J. W. - 1966
    Structure and probable growth of the oceanic business:
    Undersea Technology, VII, 5, 66-67.

Clawson, M., Landsberg, H. L., and Alexander, L. T. - 1969
    Desalted seawater for agriculture, is it economic?:
    Science, 164, 1141.

Clift, A. D. - 1967
    North Sea oil and gas exploration: Oceanology International,
    II, 1, 28-31.

Crosby, D. G. - 1970
Mineral resources activities in the Canadian Off-shore:
Maritime Sediments, 6, 30-33.

Cole, B. J. - 1974
Planning for shoreline and water uses: Kingston, University
of Rhode Island.

Cruishanck, M. J. et al. - 1968
Offshore mining: present and future: Engineering and Mining
Journal, 169, 1, 84-91.

Degens, E. T., Ross, A., and Hunt, J. M. - 1967
Red Sea: detailed survey of hot brines areas: Science,
156, 514-516.

Doumani, G. A. - 1973
Ocean wealth: policy and potential: New York, Spartan.

Durante, R. W. - 1967
Economic desalting methods: Ocean Industry, II, 5, 70-72.

Ehrlich, P. and Ehrlich, A. - 1970
The food-from-the-sea-myth: Saturday Review, April 4.

Ewing, M., Horn, H., Sullivan, L., Aitken, T., and Thorndike, E.
1974
Photographing manganese nodules on the ocean floor:
Oceanology International, VI, 12, 26-32.

Firth, F. E. (Ed.) - 1969
The encyclopedia of marine resources: New York, Van Nostrand
Rheinhold.

Fye, P. M. - 1966
The economic potential of the oceans: Industry, XXXI, 8 & 11,
etc.

Gaskell, T. F. (Ed.) - 1971
Using the oceans: London, Queen Anne Press

Gerard, R. D. and Roels, O. A. - 1970
Deep ocean water as a resource for combined mariculture,
power and fresh water production: Marine Technology Society
Journal, IV, 5, 69-79.

Gomella, C. - 1966
La soif du monde et le dessalement des eaux: Paris, Colin.

Goodier, J. L. and Sochle, S. - 1971
    Protecting the environment during marine mining operations:
    Oceanology International, VI, II, 25-27

Gordon, H. S. - 1954
    The economic theory of a common property resource: Journal
    of Political Economy, 62, 124-142.

Gordon, B. L. - 1970
    Man and the sea: New York, Doubleday

Gordon, B. L. - 1974
    Marine Resources Perspectives: Watch Hill, R.I. , Book and
    Tackle (Northwestern Univ., Earth Sci. Dept.).

Gordon, B. L. - 1977
    Secret lives of fishes: New York, Grosset & Dunlap

Gordon, B. L. - 1977
    Energy from the sea: Watch Hill, R.I., Book & Tackle Shop

Gray, T. J. and Gashies, O. K. (Ed.) - 1971
    Tidal power: New York, Plenum Press

Gruber, M. - 1968
    The healing sea: Sea Frontiers, 14, 2, 74-76.

Hargis, W. J. and Laird, B. L. - 1971
    The environmental, resource-use and management needs of the
    coastal zone: Gloucester Point (Va.), Virginia Institute of
    Marine Science.

Haulot, A. - 1974
    Tourisme et environment: Verviers, Marabout.

Hayes, M. O. - 1969
    Coastal environments: Amherst, Univ. of Massachusetts.

Helms, J. W. - 1970
    Rapid measurement of organic pollution by total organic car-
    bon and comparisons with other techniques: open file report:
    Menlo Park, California, U.S.G.S., Water Resources Division.

Hill, M. - 1975
    Diving and digging for gold: Healdsburg (Cal.), Naturegraph
    Publ.

Hillman, R. E. - 1969
    Drugs from the sea: Oceanology International, II, 1, 33-37

Holt, S. G. - 1969
    The food resources of the oceans: Scientific American,
    221, 3, 178-194.

Hood, D. W. (Ed.)
    Impingement of man on the oceans: New York, Wiley.

Hooper, M. W. - 1971
    Disposal of wastes from vessels: Oceanology International,
    VI, 9, 37, 38.

Hoult, D. P. (Ed.) - 1969
    Oil on the sea: New York, Plenum Press.

Iversen, E. S. - 1967
    Farming the sea: Oceanology International, 2, 4, 28-30.

Johnson, J. J. - 1968
    Oceanography's role in developing marine resources: Commer-
    cial Fisheries Review, March, 27-38.

Kash, D. E., et al. - 1973
    Energy under the oceans: a technology assessment of outer
    continental shelf oil and gas operations: Norman, (Okla.),
    Univ. of Oklahoma Press.

Ketchum, B. H. (Ed.) - 1972
    The water's edge: critical problems of the coastal zone:
    Cambridge, Mass., Mass. Inst. Techn. Press.

Kinne, O. (Ed.) - 1970
    Marine ecology (vol. I, Part I: Environmental factors)
    New York, Wiley.

Klima, E. F., et al. - 1971
    Attraction of coastal pelagic fishes with artificial struc-
    tures: (Miss.) Nat. Marine Fisheries Service.

Knauss, J. A. - 1973
    Ocean pollution: time to worry: U.S. Info. Serv. Impact,
    4, 18-20.

Knott, R. C. - 1975
    Who owns the oceans?: Proc. U.S. Nav. Inst., 100, 3, 853,
    66-71.

Krieger, D. - 1974
    Nuclear energy and the oceans: Proc. Pacem in Maribus IV.

Latortue, G. - 1970
    The demand for water based recreation in Southwest Puerto
    Rico: Mayaguez (P.R.), Univ. of.

Lehmann, E. J. - 1974
    Ocean waste disposal: A bibliography with abstracts:
    Springfield (Va.), Nat. Techn. Inform. Serv.

Lynch, E. J.,Doherty, R. M., and Draheim, G. P. - 1961
    The groundfish industries of New England: Washington, U.S.
    Fish and Wildlife Service.

MacDonald, G. J. F. - 1967
    What's in the ocean?:   International Science & Technology,
    April, 34-48.

MacLean, W. B. - 1966
    Future exploration of the ocean: Astronautics and Aeronau-
    tics, IV, 8, 74-76

Margolis, S. V. and Burns, R. G. - 1976
    Pacific deep-sea nodules: their distribution, composition,
    and origin: In: Donath et al., Annual Review of Earth and
    Planetary Sciences (Vol. IV), Palo Alto (Calif.), Annual
    Review Inc.

Marine Technology Society - 1966
    Exploiting the ocean (with supplement): Washington, M. T.
    Soc.

Martin, C. - 1971
    Formulating an oceanic jurisprudence: Miami, Fla., University
    of Miami Sea Grant Program, (Sea Grant Special Bulletin No. 1).

McKelvey, V. E. and Wang, F. F. H. - 1969
    World subsea mineral resources: Misc. Geol. Investigations
    Map I-632.

Mero, J. L. - 1966
    The mineral resources of the sea: New York, Elsevier.

Moore, J. J. - 1968
    The oceans. An industrial and economic perspective: Journal
    of Ocean Technology II, 3 (Oct.) 121-125.

N.A.M. - 1971
    Oceans, the new frontier: New York, National Association of
    Manufacturers.

NEMRIP - 1975
    Pacific salmon flourishing in Maine: New Engl. Mar. Res.
    Info. Progr., Information, 76, 1-2.

N.S.F. - 1973
    Inter-university program of research on ferromanganese depo-
    sits of the ocean floor. Phase I report: Washington, Na-
    tional Science Foundation.

Othmer, D. F. and Roels, O. A. - 1973
    Power, fresh water, and food from cold, deep sea water:
    Science, vol. 182, 12, October, 1973, 121-125.

Pell, C. - 1969
    The oceans - Man's last great resource: Saturday Review, 11.

Penzias, W. and Joodman, M. W. - 1971
    Man beneath the sea. A review of underwater ocean engineer-
    ing: New York, Wiley.

Pratt, S. D., Saila, S. B., Gaines, A. G., Jr. and Krout, J. E. -
1973
    Biological effects of ocean disposal of solid waste: Kings-
    ton, Univ. of Rhode Island (Mar. Tech. Rep. Series No. 9).

Report - 1969
    Our nation and the sea: Washington, Commission on Marine
    Science, Engineering and Resources, 49-82.

Report. - 1975
    Mining in the continental shelf and in the deep ocean:
    Washington, D. C., National Academy of Sciences.

Report - 1975
    Petroleum in the marine environment: Washington, National
    Academy of Sciences.

Robinson, M. A., and Crispoldi, A. - 1975
    Living resources of the oceans: Oceanus 12, 2.

Ross, F. X. - 1970
    Undersea vehicles and habitats: the peaceful uses of the
    ocean: New York, Crowell.

Roux, T. Braconnot, J. L. - 1974
    L'homme et la pollution des mers: Paris, Payot.

Ryther, J. H. - 1975
    How much protein and for whom: Oceanus 18, 2.

Salle, M, and Capestan, A. - 1957
    Travaux anciens et recents sur l'energie thermique des mers:
    (Quatrie. journees de l'Hydraulique), La Houille Blanche,
    II, 702-711.

Schaefer, M. B. - 1965
    The potential harvest of the sea: Transactions of the Amer-
    ican Fisheries Society, 94, 2, 123-128.

Scott, F. and Scott, W. - 1970
    Exploring ocean frontiers: a background book on who owns
    the seas: New York, Parent's Magazine Press.

Sette, O. E. - 1966
    Ocean environment and fish distribution and abundance, Ex-
    ploiting the Ocean: Transactions of the 2nd Annual Marine
    Technology Society Conference and Exhibit, 309-318.

Sevette, P. - 1958
    L'énergie dans les pays en voie de developpement: Paris,
    Comm. Econom. pour l'Europe des Nat. Unies.

Shigley, C. M. - 1951
    Minerals from the sea: Journ. of Metals, (Jan.) 3-7.

Sibthorp, M. M. - 1969
    Oceanic pollution: a survey and some suggestions for control:
    London, The Davies Memorial Institute of International Studies.

Simpson, M. R. - 1968
    Economic development and progress in oceanography: a
    bibliography: Atlanta (Ga.) Georgia Institute of Technology.

Smith, D. T. - 1967
    The sea floor. A new El Dorado?: The Advancement of Science,
    Vol. 24, Sept., 107-118.

Taylor, D. M. - 1971
    Worthless nodules become valuable: Ocean Industry, VI, 6, 27.

Taylor, F. B. - 1967
    Outlook for shallow oil exploration and development, United
    States: Am. Assoc. Petrol. Geol. Bull., 51, 1, 134-141.

Tooms, J. S. - 1970
    Metal deposits in the Red Sea: Underwater Science and
    Technology Journal, 28, 29.

Turvey, R. and Wiseman, J. (Ed.) - 1957
    The economics of fisheries: Rome, Food & Agr. Org.

Vaillant, J. R. - 1969
    Les problèmes du dessalement des eaux de mer et des eaux
    saumatres: Paris, Eyrolles.

Vigneaux, M. - 1971
    Potentialités adaptives de la region aquitaine: Dossiers
    du Comité d'Expansion Aquitaine, 5, (Mai.), 3-16.

Vines, W. R. - 1970
    Recreation and open space: West Palm Beach. (Fla.). County
    Area Planning Board.

Walton, W. C. - 1970
    The world of water: New York, Taplinger Publishing Co.

Weeden, S. L. - 1975
    Ocean thermal power: Ocean Industry, 10, 9, 219-228.

Weinberg, B. - 1967
    Fish protein concentrate: Present Status and future poten-
    tial: Fishing News International, VI, I, 16-22.

Wenk, E., Jr. - 1969
    The physical resources of the ocean: Scientific American,
    221, 3, 167-177.

Wenk, E., Jr. - 1969
    Seminar on the multiple use of the coastal zone (Williams-
    burg, Va.): Washington, Nat. Council on Marine Resources and
    Engineering Development.

Wenk, E., Jr. - 1972
    The politics of the ocean: Seattle, Univ. of Washington Press.

Zaalberg, P. H. A. - 1970
    Offshore tin dredging in Indonesia: London, Institute of
    Mining and Metallurgy.

Index

CREDITS
Diagrams and Photos

Page XVI: National Energy Plan, Executive Office
  of the President
Page 10: Hubbard Scientific Company, Northbrook, IL
Pages 16-19: Martek Instruments, Inc., Irvine, CA
Pages 20, 21: Center for Marine and Environmental
  Studies, Lehigh University, Bethlehem, PA
Pages 24, 26: Maritime Safety Division, U.S. Naval
  Oceanographic Office
Page 32: Fish and Wildlife Service, U.S. Department
  of Interior
Pages 39, 40, 41, 67, 75, 78: Roger H. Charlier
Page 42: OCEANEXPO-75
Page 45: EDF Service Creation-Diffusion, Paris, France
Page 59: Ancient Egyptian Fishing by Oric Bates
Page 64: Bernard L. Gordon
Page 68: T. Stoee Fotograph, Holland
Page 77: Guy Real
Page 84: Miami Seaquarium
Pages 97, 98: Lamont-Doherty Geological Observatory
Page 117: Marine Sciences Council
Page 118: C.G. Doris, Box 3010, 4030 Hinna, Norway
Page 126: FOTO KLM Aerocarto N.V.
Pages 132, 157: U.S. Navy
Page 142: Bureau of Land Management, U.S. Department
  of Interior

ABOUT THE AUTHORS

ROGER H. CHARLIER is a Professor of Oceanography, Department of Geography, Northeastern Illinois University, Chicago. He did his undergraduate work in geology, geography and political science in Belgium, and has a Ph.D. in geoscience from Bavaria, a Litt.D. in geography and a Sc.D. in oceanography from the University of Paris. He has done post doctoral work in Canada and the United States. Dr. Charlier was recently awarded a knighthood in the Order of Academic Palms by the President of France in recognition of his contributions to scientific research and international cooperation.

BERNARD L. GORDON has been a member of the faculty of Northeastern University since 1961, teaching oceanography and earth sciences. Previous to this post he served as a teaching fellow at Boston University and Instructor at Rhode Island College. He has done research on the marine fisheries of Rhode Island and oceanographic history and marine resources. Professor Gordon has published over one hundred articles and papers and is the author of many books on marine subjects and natural history.